Computer Science Workbench

Editor: Tosiyasu L. Kunii

H. Toriya H. Chiyokura (Eds.)

3D CAD
Principles and Applications

Translated from the Japanese by Hiromi Yaguchi

With 177 Figures, 8 Colour Plates and 9 Tables

Springer-Verlag
Berlin Heidelberg New York
London Paris Tokyo
Hong Kong Barcelona
Budapest

Dr. Hiroshi Toriya
RICOH Company, Ltd., Software Division
Tomin-Nissei-Kasugacho-Bldg.
1-1-17, Koishikawa
Bunkyo-ku Tokyo, 117, Japan

Prof. Dr. Hiroaki Chiyokura
Faculty of Environmental Information, Keio University
5322, Endoh
Fujisawa-shi Kanagawa, 252, Japan

Originally published in Japanese as by
Kyoritsu Shuppan Co., Ltd., Tokyo, Japan

ISBN 3-540-56507-8 Springer-Verlag Berlin Heidelberg New York
ISBN 0-387-56507-8 Springer-Verlag New York Berlin Heidelberg

Library of Congress Cataloging-in-Publication Data
3-jigen CAD no kiso to ōyō. English: 3D CAD principles and applications / H. Toriya, H. Chiyokura, eds. ; translated from the Japanese by Hiromi Yaguchi. -- (Computer science workbench) Translation of: 3-jigen CAD no kiso ōyō. Includes bibliographical references and index. ISBN 0-387-56507-8
1. Computer graphics. 2. Designbase. 3. CAD/CAM systems. 4. Three-dimensional display systems. I. Toriya, H. (Hiroshi), 1960- . II. Chiyokura, Hiroaki. III. Title. IV. Series.
T385.A1513 1993 620'.0042'0285--dc20

Typesetting: Camera ready by author

33/3140 - 5 4 3 2 1 0 - Printed on acid-free paper

PREFACE

In recent years, 3-dimensional computer aided design (3D CAD) has been recognized as one of the most important technologies for survival of enterprises. Conventionally, products are designed and manufactured through 2-dimensional data media, so-called drawing. At present, however, drawings are giving way to the media based on 3-dimensional data processed in a computer. It is not too much to say that use of the 3-dimensional CAD systems and their application techniques can be the key to winning the competition in the development of new products. Since 1984, RICOH Software Division has been independently researching and developing DESIGNBASE, a 3-dimensional CAD system. At present, DESIGNBASE is widely used as a practical system in design and production centers. This book introduces the implementation method of a 3-dimensional CAD system and its wide range of applications, based on the development of DESIGNBASE Version 3.

DESIGNBASE originates with my thesis for the master's degree at Keio University. When I was a student in the master's course of the graduate school, I read a paper about curved surfaces by the British mathematician Gregory. I still remember that the paper deeply impressed me at that time. I was also convinced that the idea would greatly influence the future CAD systems. After that, I went on to pursue a doctorate in precision machinery engineering at the University of Tokyo. There I developed MODIF as an experimental system in which the functions of the Gregory patch, my original surface, could be fully used. The concept of this system would later be applied to DESIGNBASE.

The development of DESIGNBASE started in 1984 in RICOH Software Division and Version 1 was completed in 1987. Version 1 was immature and many problems regarding its practical use were pointed out by the users. Later on, these problems were gradually solved and the system improved with the development of Versions 2 and 3. Up till now DESIGNBASE has been offered to many users. In Versions 1 and 2, the Gregory patch was the only surface supported in DESIGNBASE, whereas in Version 3, many other types of surfaces, such as the quadric surface, and the rational Bézier surface, were added. Nonetheless the Gregory patch is constantly playing the principal role in DESIGNBASE.

The authors of this book are the developers of DESIGNBASE Version 3 and are listed below by chapters. All are staff members of RICOH except Takashi Nagafuku and Fuminori Gotoh of KUBOTA Corporation.

Chapter 1	Hiroshi Toriya, Kenji Ueda
Chapter 2	Hiroshi Toriya
Chapter 3	Tsuyoshi Harada, Hiroshi Toriya
Chapter 4	Miki Satoh, Teiji Takamura
Chapter 5	Koichi Konno, Teiji Takamura
Chapter 6	Teiji Takamura, Satoshi Higuchiuchi
Chapter 7	Toshiaki Satoh, Hiroshi Toriya
Chapter 8	Toshiaki Satoh
Chapter 9	Tsuyoshi Harada, Hiroshi Toriya
Chapter 10	Kenji Ueda, Takashi Hashimoto
Chapter 11	Nobuko Fujisawa
Chapter 12	Hideki Tamura
Chapter 13	Hideki Tamura
Chapter 14	Takashi Nagafuku, Fuminori Gotoh (Kubota Corporation), Masanori Harada
Translation:	Hiromi Yaguchi

Because of a limitation of pages, some explanations may not be sufficiently thorough. For those parts, the bibliography is presented at the end of this book, which can be referenced through superscripts in the text. Use these references for deeper understandings.

At present, DESIGNBASE Version 4 is released from RICOH COMPANY, LTD. In this current version, NURBS is introduced as one of the basic representation methods of surfaces. The integration of surface and solid models, which is highly appreciated in the practical environment, is also realized. This is not our goal, and we are planning to keep on enhancing the functions.

Finally, I would like to thank Dr. Hideko S. Kunii, a director of the Software Division, for a lot of advice for writing this book. I also sincerely thank Nobuo Yamada of KUBOTA COMPUTER INC. and Tadao Itoh of D-MEC LTD. for their help.

My sincere appreciation goes to Motoyasu Sasaki and Makiko Sugawara for setting up and operating the software environment to edit this book on computer.

My thanks are also due to Yoshinobu Satake, Kazunari Kojima, Shinichiro Tsukimura, Masaaki Kagawa, Tsukasa Matsuoka, Hung-tien Ting, Manabu Tsumagari, Teruaki Itoh, Masanori Makino, Kenji Kohri, Ikumi Inoguchi, Ichiro Kamiyama, Jun Satoh, Akihiro Miyazawa, Rieko Sakamoto, Tohru Fukuhara, Koji Aragaki, Koji Tanaka, Shusaku Furushima, Naoyuki Satoh, Chikara Kagimoto, Katsumi Awaji, Masahiro Nagase, Chatman Wong, and Kenji Iino, for their assistance in developing DESIGNBASE Versions 3 and 4.

Tokyo, February, 1993
Hiroaki Chiyokura

Table of Contents

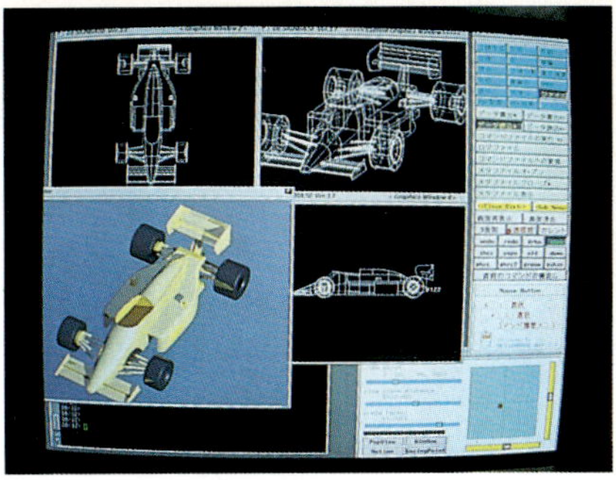

Plate 1 DESIGNBASE
Version 3 Garphical User Interface

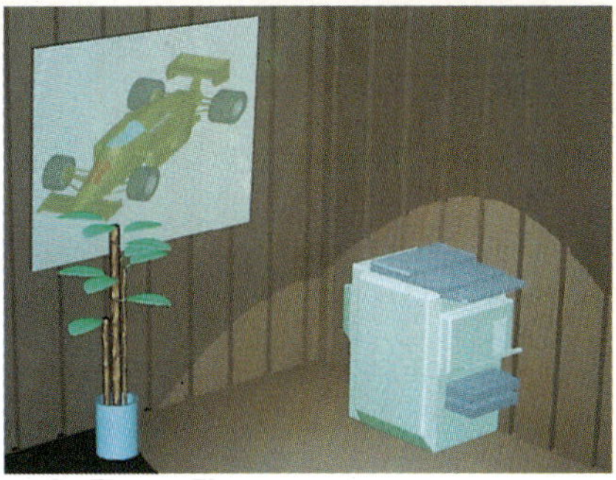

Plate 2 Example of Designing a Copy Machine

Plate 3 Shaded Image of Gourourd Shading Method

Plate 4 Shaded Image of Scan Line Method

Plate 5 Shaded Image of Ray Tracing Method

Plate 6 Resin Models Created
by Laser Lithographic System

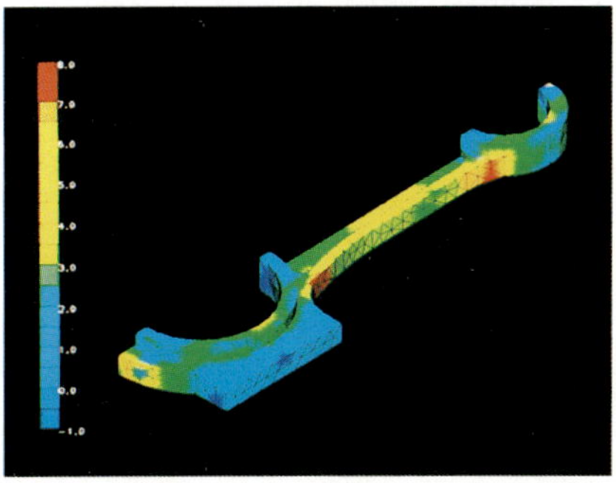

Plate 7 Result of Stress Analysis of a Connecting Rod

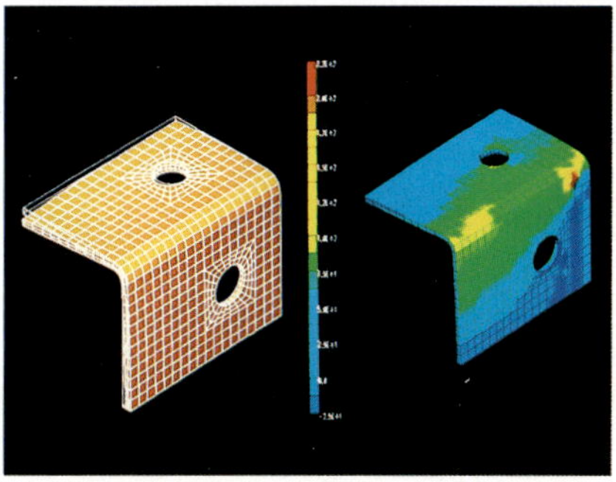

Plate 8 Result of Stress Analysis of a Frame

1. OUTLINE OF CAD/CAM/CAE

1.1 Introduction

At present, consumer's various needs must quickly be responded to and shortening the cycle from the product planning stage to the design and manufacturing stage has become the major function of manufacturers who want to succeed in a competitive environment. One of the methods currently suggested for shortening the cycle of these procedures is concurrent engineering, which involves designing and manufacturing in parallel. The most powerful tool that can be used to make the design and manufacturing process of a product effective is a computer. **CAD** (Computer Aided Design) and **CAM** (Computer Aided Manufacturing) have become popular as tools essential for designing and manufacturing with computers. **CAE** (Computer Aided Engineering) is also becoming popular, for verifying products or determining whether their parts can actually be used, by using computers. This chapter explains the outline of CAD/CAM/CAE systems and how they have been developed. As these systems must be organized to design and manufacture products efficiently, the present condition of the systems is also described.

1.2 CAD System

CAD allows the use of **computer graphics** (CG) in the field of design. The CAD project of the Massachusetts Institute of Technology (MIT) in 1959 may have been the first full-scale project to develop the system that processes drawings in computers. This project had been completed in the early 1960's by Sutherland and had been presented as the SKETCHPAD[140] system. Although the SKETCHPAD system was an early drawing processing system that used computers, the system included most of the concepts that can be seen in today's interactive 2-dimensional CAD system. At that time, General Motors developed a CAD system for designing automobiles called DAC-1. As a business system, CADAM of Lockheed Corp. had been put to practical use in 1967.

Business systems in their early stages were used by connecting many graphics terminals to a large computer. This created problems because much burden was imposed on the large computer and the interface between terminals and the large computer occurred too often. To solve these problems, an intelligent-type terminal was developed, so that high-speed drawings would be realized on terminals with microcomputers being incorporated in the terminals. With this system, the burden on the large computer and the interface between terminals and the large computer were reduced. Hence, the speed of response and operation of the system were dramatically improved. This type of the system has since been introduced in many enterprises which employ a lot of designers.

At the same time, the turn-key type system which incorporated minicomputer in CAD systems had also been developed. In 1968, Applicon Corp. introduced AGS, and in 1969, Computer Vision Corp. introduced CADDS. These systems were upgraded through connecting a mini-computer to a dedicated graphics terminal. As it was very easy to introduce this type of system for use, it became popular in the late 1970's. After this, Autodesk Corp. introduced AutoCAD, its 2-dimensional CAD system operated on personal computers. This system provided users 80 percent of the functions of large-scale computers, but at lower costs, and therefore sold in great numbers. The success of AutoCAD largely contributed to the spread of CAD systems. In recent years, a personal computer called **engineering workstation** (**EWS**) has become increasingly popular. Some engineering workstations can display 3-dimensional figures very quickly and some can perform calculations as super computers do. The systems developed so far are to be ported to the workstation (WS). New CAD systems with the improved characteristics of workstations are being developed one after another.

As new CAD systems were being developed, the necessity to standardize the graphics system was recognized, and in 1984, the **Graphical Kernel System** (**GKS**) [73,2,52,67] was established by International Standards Organization (ISO). However the GKS had one weak point, in that 3-dimensional figures could not be represented easily, and so in 1987, the GKS-3D [74] in which the weak point was solved was standardized. Also, in 1989, **Programmer's Hierarchical Interactive Graphics Standard** (**PHIGS**)[134,135], whose biggest characteristic is to use data hierarchically, became the ISO standard. **PHIGS+**[42] to which shading functions and regulations on light source handling are added, was standardized in early 1990's. Moreover, **PHIGS Extensions to X** (**PEX**)[122] was released in 1991. In this system, X Window System, which is the standard of the window system and PHIGS+ are synthesized. If 3-dimensional displays are realized in the X Window System, the software development of CAD with using this system might be greatly affected. In addition, the graphics libraries such as the **Open GL** developed by Silicon Graphics Inc. and **HOOPS** developed by Ithaca Software, are competing in the market to be the standard. Generally speaking, systems have been standardized through integrating technologies spread in markets. However, it is dangerous just to follow technologies in the CAD and CG fields because the standard does not always meet the actual situation. Therefore, the technologies must be anticipated as much as possible, then the standard established.

Observing software technologies of CAD, opened up the field of computer aided geometric design (CAGD) , which is the system to construct 3-dimensional shapes in computers. With this the technologies of solid modeling to represent 3-dimensional shapes completely and the technologies to design complicated shapes containing free-form curves or surfaces have accumulated. Since the computer has become able to represent complicated shapes, the technique to display those shapes beautifully, called **rendering**, has emerged as one of the important fields in computer graphics. Several rendering methods have been researched and developed: scanline method, ray tracing method, or radiosity method. With these methods, images can be obtained that resemble real objects.

The technologies of CG and CAD have been developed using technologies of software and hardware interacting with each other from the beginning. At the early stage, what is important for hardware technology is to input and output elements of drawings and to process designs interactively. But recently, the technology used in computer graphics to effect mathematical techniques as quickly as possible is required for hardware. To meet this requirement, a workstation to rotate and shade 3-dimensional shapes simultaneously or a system to realize various rendering method on hardware are gradually being popularized. Such competition of software technology and hardware technology is one of the characteristics in the CAD field.

1.3 CAM System

CAM is the general name of the technology used to apply computers in the manufacturing of industrial products. The CAM system is usually coupled with the CAD system, and referred to as the "CAD/CAM system". It has been regarded as an integrated system to deal with the processing steps of a product from design to manufacturing. The data designed by a CAD system is used up to the manufacturing process, from process planning, work division and tool preparation to acquiring production and test data. Systems to support design and manufacturing by computers have been developed for many years, but such systems cannot completely support the CAM system at present. An integrated CAD/CAM system has not yet been realized in a real sense.

CAM has a longer history than that of CAD. It began in 1949 with the MIT's development of a numerical control (NC) machine. Numerical control is a technique to control machines by numeral operations. In 1952, **Automatically Programmed Tools** (**APT**)[21], a software to describe such operations, was developed. APT can process 3-dimensional data and many expanded software programs have been developed so far. Later on, APT was expanded in its function in West Germany and **Extended subset of APT** (**EXAPT**)[139] was developed in the 1960's. In EXAPT, the conditions for reducing processing can be semiautomatically determined. The functions of APT have now been expanded further with CAM-I(Computer Aided Manufacturing-International) which was established in 1971.

As explained so far, applications of computers to design and manufacturing are not fully utilized as they have not yet reached to the stage of practical use. The following could be factors that prevent an integrated CAD/CAM system from being constructed:

- Choice of industrial tools, processing conditions and procedures are not completely automated.
- An integrated database to deal with design and manufacturing data has not yet been constructed. Neither has the method to create databases from accumulated know-how been established.
- Interface for exchanging data between systems has not yet been standardized, and data cannot be correctly sent and received between CAD/CAM systems.

Let us now examine the present conditions of interface between systems. At present, **Initial Graphics Exchange Specification (IGES)**[136] is most popular as the standardized interface between CAD/CAM systems. IGES specifies the format for exchanging drawing data, and most of CAD/CAM systems can exchange data through the IGES format. However, the way to represent curves and surfaces are different in each system, and sometimes such shapes need to be approximated in order to output them according to the IGES rules. There is another problem; although a surface is expressed with the IGES specifications, few systems can correctly read the data. If a surface expression is complicated, the less the system is able to read the data. To compensate this situation, IGES has been improved with the progress of technology by introducing NURBS, which can represent surface expressions quite well, or by incorporating solid data representation. In the IGES Version 5.0, application support will be expanded with an international view, to include Chinese characters. Also, STEP [75] is now being examined as the next standard of IGES and problems on interface are being gradually solved.

What is important to develop an integrated CAD/CAM system is the standardization of interface and decentralization of processing. The work sharing can be realized by enhancing personal computers and workstations, and the protocol that connects these terminals with the network is now being examined. An integrated CAD/CAM system can only be accomplished in the true sense when data can be interfaced between the personal computers and the workstations. The incorporation of know-how and information is under examination to construct a database for the design and manufacturing processes and to develop a system for automatic schedule layout.

1.4 CAE System

CAE can be defined as a system engineering approach to the product development processes such as design and manufacturing. In other words, CAE aims to check validity of design during product development and to reduce man-days of manufacturing, by applying mathematical models to the design and manufacturing stages. CAE is usually divided into two category: electrical CAE system and mechanical CAE system.

Electrical CAE system is used to design semiconductor circuits such as LSI or VLSI or to design systems. Computers are used to design or simulate logic circuits or to automatically design the layout of printed boards. LSI is highly integrated at present, so no products can be developed without computers. In the 1970's, the CAD system to design the layout of LSI was developed, and in the 1980's, the CAE system to check the result of the layout through simulation was introduced. At present, a system known as **silicon compiler**[71] which automatically determines the layout of chips by inputting logical specifications is about to be put into practical use.

In the mechanical CAE system, computers are used to analyze, evaluate and optimize products at the design stage. To do this effectively, modeling and systems analysis are necessary. The nucleus of the modeling is solid modeling, because it can accurately represent shapes. Even though mathematical models are used to analyze the shapes, the analysis will be incorrect if the shapes are not accurate. As such, solid models can help CAE achieve good results. Systems analysis consists mainly of the following three types of analysis: structure analysis such as strength analysis or deformation analysis, thermal analysis such as heat transfer analysis or thermal stress analysis, and fluid analysis such as fluid flow analysis or resin flow analysis. Of these, structure analysis using **finite element method** is the most popular.

Since the introduction in 1974 of SUPERTAB, the analysis modeling software with a finite element method, various softwares for system analysis have been developed. The initial systems contained the following problems: vast amount of time was needed for calculations and only specialists could handle the software because the result was output with numerals. Because of these problems, in most enterprises, designers had to request specialists for analysis when using CAE systems. However, the processing speed of computers has become very fast in recent years and the post processor that visualizes the results of an analysis has become popular. Thanks to this progress, the CAE system has gradually become designer-friendly.

The CAE system can reduce the frequency of prototype creations and experiments, can shorten the development period, and can save on development costs. Also, by constantly evaluating and adjusting the design according to the analysis, the design can be optimized. In order to strengthen the connection to CAD systems, the number of CAE systems that support IGES has been increasing over the years. If designers are able to use the CAE systems with simple operations and no special knowledge, and get precise results, the CAE systems would quickly become popular in the near future.

1.5 3D CAD Systems

Many enterprises have been promoting automation of the design and manufacturing processes by introducing CAD/CAM/CAE systems to design their products and parts. As a result of its introduction, many enterprises point out shorter period of design, lower cost of design work, and improvements in product quality. On the other hand, as CAD/CAM systems become popular, requests have been made for systems which deal with the A to Z of design at the site of design and manufacturing[95]. Drawing is a method of representing shape data at the present. If 3-dimensional data of a drawing is needed for analysis and processing, the data should be directly inputted into the computer according to the drawing. This situation still requires improvement even though CAD systems have been introduced for drafting. However most of the practical CAD systems at present deal only with 2-dimensional shape data.

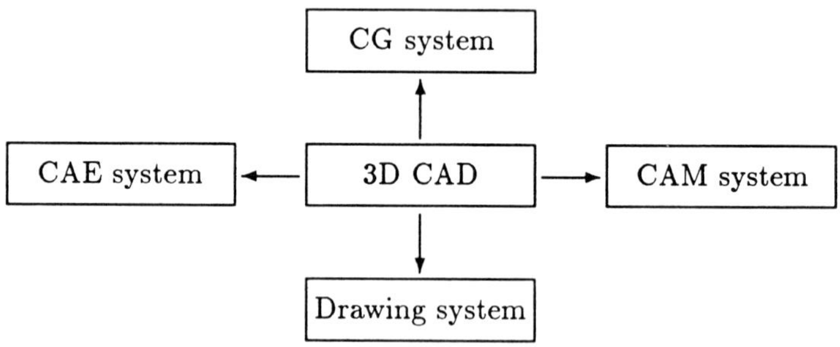

Fig. 1.1. System with 3D CAD as nucleus

Shape data should be 3-dimensional, not 2-dimensional, if it is to be used at all stages in industry, from design to manufacturing. As Figure 1.1 shows, CAD/CAM/CAE systems can be integrated in the true sense, with the CAD system (such as a solid modeler that deals with 3-dimensional data) being set at the center. This situation leads to efficient work of design and manufacturing. For example, a drawing can be obtained by sending the data in the 3D CAD system to the drafting system. If the data is passed between the 3D CAD system and the CAE system, it can be modified in the 3D CAD system according to the result given by the CAE system. The real image or animation can be created by sending the data to the CG system. If one wants to have the real product in hand, the data in the 3D CAD system can be processed in the CAM system.

How to realize the interface between these systems is still a big problem because each system stores data in different forms. For example, curved surfaces need to be represented within some range, in order to be processed in the CAM system. On the other hand, the surfaces need to be modeled with the mesh structures in the CAE system and most CG systems use polygons to represent the surfaces. The 3D CAD system itself has much room for development.

What leads the 3D CAD systems is the solid modeler. In this book, the solid modeler DESIGNBASE which is developed by the authors of this book is introduced as an actual example. This book explains the development history of the solid modeler and their wide range of applications, based on the development of DESIGNBASE, e.g., how curves and surfaces are represented in computers and how body shapes are inputted in computers. For curves and surfaces, this book also describes their mathematical representation and intersection calculation. Rendering and calculation of mass properties are explained as examples of the applications which directly process 3-dimensional data. Furthermore, how to interface with CAE systems or 3-dimensional solid processing systems are explained.

2. 3D CAD SYSTEMS AND SOLID MODELS

2.1 Introduction

The preceding chapter described how important the roles of CAD/CAM/CAE are and how they affect the design and manufacturing of products. In recent years, many products, including household electrical appliances such as telephones or tape recorders, have taken on unique and complicated surface shapes. Such shapes are difficult to represent in two dimensions, thus that is where 3-dimensional CAD systems come in. The configuration or construction of systems vary depending on the way the systems are to be used, i.e., displaying solid shapes in three dimensions, processing the solid shapes, or obtaining their volume or center of gravity.

The functions and performance of a system are affected by the representation method of solids in a computer. From the viewpoint of a user of a CAD system, getting information on how a solid is represented leads to a better understanding of the system's functionalities. From the standpoint of a developer of a CAD system, the requirements for the system provide a good judgement of how solids should be represented. Taking these points into consideration, this chapter explains how to represent solids in a computer and provides examples of solid models that have recently become popular in 3D CAD systems.

2.2 How to Represent Solids

How are objects represented in computers? To simply display 2-dimensional shapes, only geometric data such as coordinates of vertices and edge data that construct the shapes need to be stored in a computer. If the shapes are to be modified efficiently even in two dimensions, topological data i.e. which edge connects to which edge is further needed. In order to represent 3-dimensional objects, the surfaces need to be taken into consideration in addition to the edges and the vertices. If the surfaces are curved, geometric representation will be complicated. The topological data to indicate how the elements relate will also be complicated. Such objects can be represented by using one of the following three models: wireframe model, surface model, or solid model. Let us look at these models one by one.

- **Wireframe model**

  The wireframe model represents a 3-dimensional object only with edges and vertices. One can visualize such a structure as consisting of pieces of wire. The edges are either straight or curved and have the coordinates of their start and end points. These structures can easily be represented in a computer, because only a small amount of memory capacity is necessary and data can be accessed quickly. Therefore most commercial CAD systems have the wireframe model as a basic function. It is good enough to use the wireframe model for drawing 3-dimensional objects. The wireframe model is also used to input data for analysis by the finite element method.

  However, an object represented by the wireframe model cannot be interpreted uniquely. As shown in Figure 2.1, the wireframe model of (a) can be interpreted as either models (b), (c), or (d). Furthermore, the wireframe model does not contain surface data. In other words, this model is incomplete as an object. For this reason, the wireframe model is useless for obtaining volume, surface area, or center of gravity data, and it is impossible to display an object with its hidden lines erased automatically.

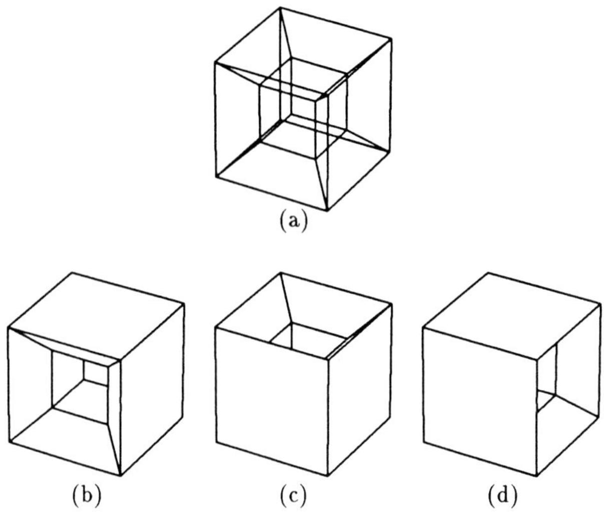

Fig. 2.1. A wireframe model that cannot be interpreted uniquely

- **Surface model**

  The surface model has surface data in addition to wireframe data, just like pieces of paper attached on to a wirework. The surface model does not contain data for its contents and is therefore void. As a topological structure, a surface also has constructing edge data. The mathematical representation of the surface data differs depending on the system. Some systems can represent free-form surfaces such as Coons or Bézier surfaces or NonUniform Rational B-Spline surfaces (NURBS), as well as natural quadric surfaces such as cylindrical or spherical surfaces. Since the cutter path for a numerical control (NC) machine can be created if a surface is defined, most of the existing 3-dimensional CAM systems are based on the surface model. Some systems can automatically calculate volume or center of gravity about an object by defining some region enclosed within the curved surfaces.

However, the surface model of an object is merely a group of surfaces, so it is not best-suited for representing solid objects. If a curved surface is split, for example, the cross-section must be intentionally reconstructed. Moreover, some space gap may exist between surfaces in which case the cutter path cannot be correctly created. Most of the CAM systems based on surface models compensate for this problem by introducing a function that corrects the cutter path.

- **Solid model**

  The solid model can completely represent an object. The contents of an object represented by the surface model are void, but the solid model can define these contents. So at the various stages of manufacturing such as analysis or processing, automatic processing can be effected. The system based on the solid model is called the **solid modeler**. Advantages of the solid modeler are explained in the next section.

  Many methods have been suggested for solid model representation in a computer. The representative methods are firstly, the method in which basic shapes called **primitives** are combined to represent one object, and secondly, the method in which an object is represented by using surface, edge, or vertex data of surface models as well as topological data.

  Compared with the wireframe and surface models, the solid model has complicated data structures and it takes a longer time to modify or display the shapes. Solutions to these problems are being studied at present.

As described in this section, the only 3-dimensional model that can completely represent an object is the solid model. Thanks to the dramatic progress in computer technology, the processing speed of a solid modeler is gradually improving. Therefore, the number of 3-dimensional CAD systems which are based on the solid modeler is increasing.

2.3 Advantages of a Solid Modeler

There are various purposes for using CAD/CAM/CAE systems: drawing blueprints, calculating volume or center of gravity, analyzing structures using the finite element method, simulating various stages in manufacturing such as processing and assembling, creating cutter paths, or displaying a realistic image with computer graphics. Some of these purposes can be realized by using the wireframe model or the surface model. These models, however, do not represent objects completely, and as such require man power during most of the processing procedures.

Actually, when the shape of a product or a machine part is inputted in a computer using a CAD/CAM system, it becomes very complicated. Furthermore, depending on the shape, it is nearly impossible to create processing data or mesh data by manual work. The solid model, however, can process data automatically even though the processing takes a long time. This is the biggest advantage of using the solid model. Other advantages with examples of solid model applications are described below:

- **Interference check**

  A machine is designed by assembling a lot of parts. Needless to say, checking the interference between the parts is quite an important process at the design stage. If the parts are represented by the solid model, interference between them can automatically be checked[16]. If necessary, the interfering area can be displayed on the screen. In recent years, some enterprises have introduced solid modelers for large-scale design. This trend[77] has indicated that the interference check remarkably saves time in designing. Errors in designing have also sharply decreased, with the introduction of the interference check function. This function can also be applied to control the movement of an industrial robot[109]. Although the movement of the robot is usually controlled by a computer, the robot may actually touch surrounding machines when it moves. If a solid modeler is incorporated in the system that controls the robot, the relationship between the robot's movement and its environment can be checked in advance. Situation in which the robot makes contact with the surrounding machines is automatically checked and as such gross accidents can be prevented.

- **Mass property calculation**

  Conventionally, the mass properties of an object (volume, center of gravity, surface area, and moment of inertia) have been calculated manually from blueprints, or by measuring a prototype. If an object is represented with the solid model, these mass properties can automatically be calculated[84,85,88]. This ability to obtain the mass properties easily during the design stage helps designers to a great extent.

- **Cutter path generation**

The cutter path is indispensable for the processing of NC machines and many methods have been proposed to obtain it automatically[61,138, 163]. A solid model has enough data to create the cutter path; by using a cutter path in a solid modeler, processing procedures can be simulated[79]. In recent years, a system has been developed, which displays the movement of a cutter in real time on a high-performance 3-dimensional graphics screen. This system can check on the screen whether or not a material is cut too much or whether a cutter interferes with an NC machine. Another method has also been proposed, which automatically verifies a cutter path from solid model data[156]. The trajectory of a cutter as solid model data is used to check the interference between the parts of an NC machine and materials.

- **Object recognition**

  A study is being made to determine the position or condition of an object from images inputted by a telecamera[9]. A solid model of the object can give some description of the object condition. By comparing this description with the image data, the real condition of the object can be recognized[160]. This study is being applied to the technology used in robots for recognizing objects at the manufacturing site.

- **Structure analysis**

  When analyzing a solid model by the finite element method, mesh data should be created from the solid data. Several methods have been proposed to create mesh data automatically from the solid model[123]: one method uses boundary representations[165,166], another uses the octree structure (refer to Subsection 2.4.3)[169]. A solid modeler that contains the automatic mesh generation function is also on the market. In spite of this, it is difficult to create effective meshes for any analysis, and specialists must adjust the meshes according to their experience and knowledge in order to perform the best analysis. Recently, a system was developed which uses the P-method, in which the finite element method is improved. The most important characteristic of the P-method is that it gives satisfactory precision in analysis although the meshes are roughly divided. If such a system is connected with a solid modeler, the structure can be analyzed without the necessity of mesh creation.

- **Solid creation system**

  In recent years, a system that creates a solid resin model has been developed and put into practical use. In this system, laser beams scan a liquid light-cured resin and a solid shape is formed. In theory, a closed curve on a surface of equal altitude is generated by a CAD system. Laser beams scan the area enclosed by the curve and then the resin is cured. By slightly changing the altitude and repeating these operations, a solid shape is created. Therefore, if the data on surfaces of equal altitude is outputted using solid data, a prototype of a product can automatically be created. Conventionally, it takes a great deal of time to make a prototype, but this system, if connected with a solid modeler, will surely help to improve the development efficiency of products.

This section has described the application range of a solid modeler. We understand that the range is expanding more and more as CAD/CAM/CAE are integrated.

2.4 Representation of Objects in a Solid Modeler

How has a solid modeler, which is at last being put into practical use, been developed? This section explains how objects have been represented in a computer by examining the history of the solid modeler.

2.4.1 CSG and boundary representations

A solid modeler was at first presented in 1973 at the international conference PROLAMAT held in Budapest in Hungary. In the conference, two kinds of solid modelers were proposed: one was the system called **TIPS-1** from the University of Hokkaido[103] and the other was **BUILD** from Cambridge University[19].

TIPS-1 represents an object by defining fundamental objects called primitives and combining the primitives by Boolean operations. This method is called **constructive solid geometry (CSG)**[113,114]. As Figure 2.2 shows below, the Boolean operation is an operation from which the union, difference, and intersection of objects can be obtained. In this figure, (b) shows the operation of the union, in which two objects are combined into one, (c) shows the operation of the difference, in which one object is removed from another, and (d) shows the operation of the intersection, in which the common part of two objects is created.

In TIPS-1, the primitives are represented by half-space. Since a face F_{ij} in space splits the space, the half-space S_{ij} can be expressed as follows:

$$F_{ij}(x, y, z) \geq 0 \tag{2.1}$$

A closed space S_j can be expressed as

$$S_j = \bigcap_{i=1}^{n} S_{ij} \tag{2.2}$$

which indicates that the common part of the half-space and an object S can be expressed as

$$S = \bigcup_{j=1}^{m} S_j \tag{2.3}$$

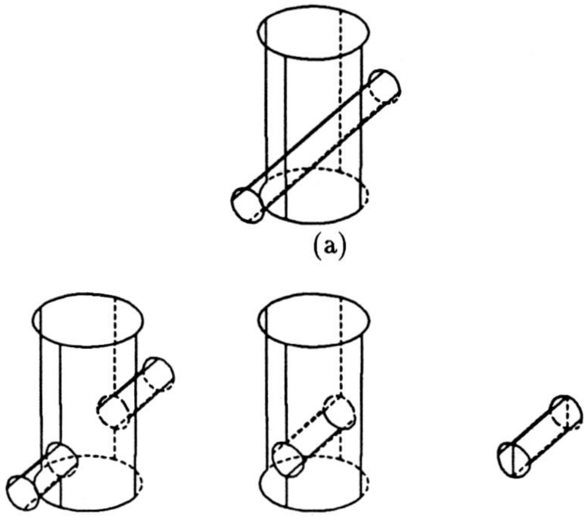
(a)

(b) (c) (d)

Fig. 2.2. Boolean operations

which in turn means the union of S_j. In the actual operation of TIPS-1, S is represented by subtracting the union of the non-existing part Q_j of the object from the union of the existing part P_i, as shown in the following expression:

$$S = \bigcup_{i=1}^{m} P_i - \bigcup_{j=1}^{n} Q_j \tag{2.4}$$

In TIPS-1, F_{ij} can be defined as a plane, natural quadric surface, or free-form surface, so that the primitive can be defined as a parallelepiped, sphere, cylinder, cone, or an object with free-form surfaces. With this system, various shapes can be represented.

In BUILD, a solid is defined by the Boolean operation of primitives as in TIPS-1. It differs from TIPS-1 however in that an object is represented by its boundary surfaces and the topological data of the surfaces is stored in the system. This method of representation is called **boundary representations**. As Figure 2.3 shows, an object consists of faces, which in turn consist of edges, and each of the edges has vertices at both its ends. The vertices also have coordinates. In this method, the system stores all the boundary data of an object thus enabling processing, such as displaying the object, to take place very quickly. The structure shown in Figure 2.3, however, has a problem in that the number of edges which comprise a surface cannot be fixed, and this is troublesome for the computer operation.

Afterward, Baumgart simplified the data structure of the boundary representations by introducing the **winged-edge structure** and solved the problem[7]. He also introduced **Euler operations** for the modification of shapes, so that the consistency of the topological structure of an object was ensured[8]. As a result, the topological structure can now be modified system-

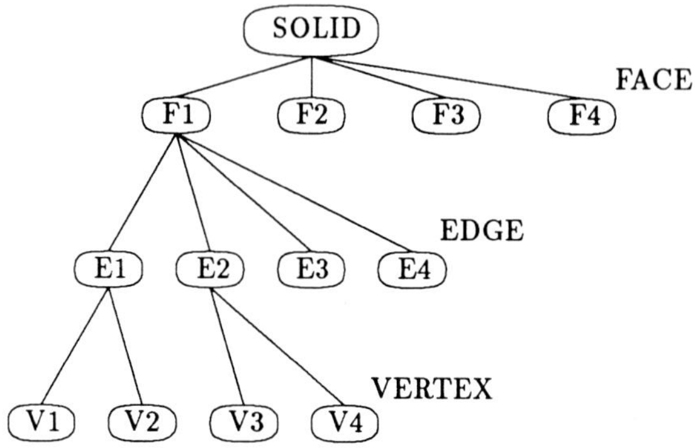

Fig. 2.3. Boundary representations

atically. Since the winged-edge structure and Euler operations are very important concepts for understanding the boundary representations, their technical aspects are explained in detail in another chapter and this section briefly explains the history of the winged-edge structure and Euler operations.

The winged-edge structure, which Baumgart proposed, could not represent an object with holes. By introducing the concept of a loop, Braid et al. made it possible to represent holes in the winged-edge structure[18]. BUILD-2 is a system that was developed with this representational method. Mäntylä further proposed the **half-edge structure**, which was an improvement on the winged-edge structure and which enabled high-speed access to shape data. Next he succeeded in developing a solid modeler GWB by using Euler operations[91, 92]. GWB showed the usefulness of Euler operations but they were not applied to all of the modification operations in GWB. In Japan, MODIF, a system whose modification operations are performed using the primitive operations of expanded Euler operations, has been realized[31,30]. DESIGNBASE which is based on the winged-edge structure and Euler operations, was also developed in Japan.

2.4.2 Hybrid systems

Voelcker and Requicha of Rochester University developed the PADL-1 system and presented its results in 1977[153]. PADL-1 is a CSG-based system for the internal representation of objects. Parallelepipeds and cylinders can be defined as primitives and an object created by combining the primitives using Boolean operations. The process involved in executing the Boolean operations is stored in the system as a **tree structure**. This tree structure regulates the order of the Boolean operations to be executed.

Figure 2.4 illustrates the tree structure of the CSG in PADL-1, in which a parallelepiped and a cylinder are combined first, then another cylinder is subtracted from the combined primitives. In PADL-1, the boundary represen-

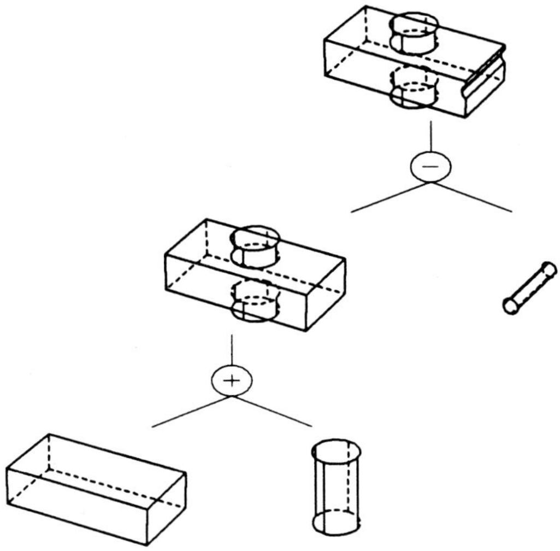

Fig. 2.4. Tree structure of CSG in PADL-1

tations are calculated from CSG data for drawing an object. A system that supports both CSG and boundary representations, as found in PADL-1, is called a **hybrid system**.

GMSolid is a system which has been influenced by PADL-1, and which uses a method that can obtain boundaries from CSG[17]. PADL-1 has now been expanded into PADL-2, by adding natural quadric surfaces as one of the primitives[20].

The major solid modelers which have been developed in recent years are listed in Table 2.1. All of the existing solid modelers use either boundary representations, CSG, or the hybrid method. This fact proves that the first boundary representations or CSG which were proposed were almost perfect.

2.4.3 Other representation methods

In addition to CSG and boundary representations, there are other methods of representing objects such as **cell decompositions** and **spatial enumeration** schemes[113]. The cell decompositions scheme represents an object by decomposing it into fundamental shapes such as tetrahedrons. Each of the decomposed shapes is called a cell. This method regards an object as a group of cells, so its volume or moment of inertia can be obtained quite easily. This method is also suitable for representing the meshes used in structure analysis.

In the spatial enumeration scheme, an object is represented by 3-dimensional arrays. Space is divided into cubes and the occupancy of the space by the cubes is indicated by 0 and 1. It is natural to represent an object with this method but the amount of data required is tremendous.

Table 2.1. The major solid modelers

System names	Suppliers	Data representations
ACIS	Spatial Technology	B-reps
Anvil 5000	MCS	CSG/B-reps
Bravo 3-SM	Applicon	CSG/B-reps
Cadds4X-SD	CV/Prime	B-reps
Catia SGM	IBM	CSG/B-reps
Cimplex-Design	CIMPLEX	CSG/B-reps
DESIGNBASE	RICOH	B-reps
Euclid-IS	Matra Datavision	CSG/B-reps
Geomod	General Electric	B-reps
Medusa 3D	CV/Prime	B-reps
ME series 30	Hewlett Packard	B-reps
ParaSolid	Electronic Data Systems	B-reps
ProENGINEER	Parametric Technology	B-reps
Tips-1	CAM-I	CSG

The **oct-tree structure**[76,96] was developed as an improvement on the spatial enumeration scheme. In this structure, the sizes of the cubes are variable and the intersection between objects can easily be found. The shape of an object, however, cannot be correctly represented because the curved-surface shapes are represented by cubes. To compensate for this problem, studies on the addition of boundary data to an object are being made[23,56]. Other studies are also being made to reduce the amount of data required by arranging the data structure[168].

If these schemes are used independently, there is still a problem in representing object shapes. However, when used with CSG or boundary representations, these methods can speed up the intersection calculations[167]. Applications that do not require so much precision, e.g., computerized axial tomography (CT) scanning used in medical treatment, have also been studied[81]. In such cases, because the data is digital, it can be represented by the octree structure without lowering precision. As indicated so far, different methods can be chosen to represent an object efficiently, depending on the required application.

2.5 Comparison of CSG and Boundary Representations

CSG and boundary representations (B-reps) which are the most popular representation methods for solid models are compared in this section. Table 2.2 shows a comparison of the two methods.

Table 2.2. Comparison of CSG and boundary representations

	CSG representations	Boundary representations
Data structure	Simple	Complicated
Amount of data	Small	Large
Validity	Represents only valid objects	Represents any objects
Data exchange	Available (into B-reps)	Difficult (into CSG)
Re-modification	Simple	Depending on data structure
Local modification	Difficult	Easy
Speed of display	Slow	Fast
Surface representation	Difficult	Relatively easy

As the table shows, with the CSG method data representation in a computer is simple, so a system can easily be constructed. On the contrary, the data structure of boundary representations is complicated, but display speed is very fast because the system has all the necessary shape data. With the CSG method, the boundary data of an object has to be created, so it takes a long time to draw an object or display its shaded image. In order to speed-up the display, an expanded CSG, in which redundant data is removed, has been proposed[121], but it has not yet been put into practical use.

Data conversion from boundary representations to CSG representations is difficult. Many studies have been made on the conversion of 2-dimensional data, and how to convert a shape on a plane into a 2-dimensional primitive[107, 154]. However, the exchange of 3-dimensional data is still being studied.

One of the advantages of CSG representations is the ease with which designs can be modified. This advantage is brought about by the characteristic of CSG representations which allows an object shape to be represented as it is as well as representing its design procedures. For example, to change the radius of the hole in Figure 2.5(a), all that is required is to change the radius of the original cylinder and to execute a Boolean operation. On the other hand, in the boundary representations only the shape data is stored, so that modification of an object shape is complicated. The hole would first need to be filled by the union operation, then a cylinder of another size subtracted by the difference operation. The Boolean operations can modify very complicated shapes but it is quite difficult to recreate the original shape after the modification. Using boundary representations, however, some systems have been able to recreate shapes by storing the information required in the design process. Some other systems have been developed which can modify shape parametrically by adding the relationship between the elements of the boundary representations.

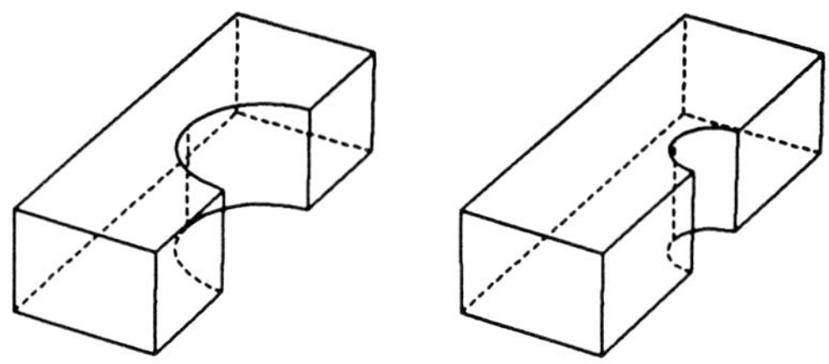

(a) (b)

Fig. 2.5. Modification of a shape with CSG representations

The biggest disadvantage of CSG representations is that the principal method for modifying shapes is the Boolean operation. Another method in which fillet surfaces are created by using offset calculations in addition to the Boolean operations has been proposed[120] but generally, local modification is hardly realized with CSG representations. The Boolean operations take a long time if they are applied to objects that have complicated shapes. To solve this problem, some systems introduce batch processing, but this cannot be applied to all situations since in most cases design work must be done interactively. Furthermore, the fact that shape modification depends on the Boolean operations in the CSG-based systems means that designing free-form surfaces is a difficult task. This is because it is still difficult to modify surfaces by using only the Boolean operations.

On the contrary, boundary representations can be used to freely modify the boundary data of objects. By modifying boundaries, e.g., moving vertices or edges and lifting surfaces, intuitive modification can be realized at high speed. This is where the boundary representation has its strong point. However, there is a disadvantage in this, in that when making modifications, an incorrect shape may be created in which an object intersects itself. In CSG representations, such intersection is prevented by the Boolean operations.

Figure 2.6 illustrates what an object that intersects itself (a self-intersecting solid) is like and how the object is created. In Figure 2.6(a), the face in bold is to be lifted. If the wrong distance is specified for lifting, the object shown in Figure 2.6(b) will be created. When this happens, there are difficulties in making the necessary corrections. The previous shape, Figure 2.6(a) should be regenerated by another modification method, but this is not easy in all

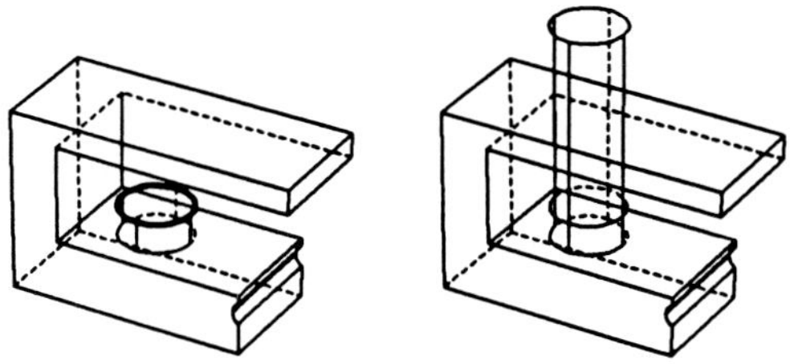

(a) (b)

Fig. 2.6. Creation of a self-intersecting object

cases. Although a system can check whether or not a self-intersecting object is created during modification, it takes a long time. However, this problem can be solved if the system stores the process of solid design in addition to the boundary representations, because then, the shapes can easily be regenerated.

There are both pros and cons associated with CSG representations and boundary representations. Each method can be used properly depending on the purpose of the system.

3. SOLID MODELER DESIGNBASE

3.1 Introduction

The previous chapter explained generally how solid modelers are developed in 3-dimensional CAD systems. This chapter introduces DESIGNBASE, a practical solid modeler, and gives an outline of its functions, module construction, data structure, and basic modification operations. DESIGNBASE is a B-reps based solid modeler; it uses primitive operations which are an expansion of the Euler operations to modify object shapes. This chapter explains how the design of an object is stored in a computer using boundary representations. The specifications of the primitive operations are also explained in this chapter.

3.2 Features of DESIGNBASE

Solid modelers have been developed to design the shape of objects. Design was carried out by using primitive shapes and repeating the Boolean operations between them. With this design procedure, however, surfaces that had complicated shapes were hard to be designed, and some were even impossible. This was one of the biggest reasons why solid modelers were prevented from spreading, even though they were indispensable in CAD, CAM, and CAE technologies.

The solid modeler DESIGNBASE was developed to deal with such problems. DESIGNBASE has its strongest point as a solid modeler in the ease of design and modification of surface shapes. This characteristic was enabled by establishing mathematical representations to precisely store the shapes of surfaces in a computer and by introducing the mechanism on how to design the shapes.

In DESIGNBASE the Gregory patch is another feature, which enables smooth connection of surfaces, so that complicated shapes have been able to be represented. The rational form of representation is also introduced in DESIGNBASE, allowing conic sections and quadric surfaces, which are significant in engineering, to be expressed precisely.

DESIGNBASE realizes surface design and modification through various functions, e.g., the substantial rounding operation which creates fillet shapes popular in CAD. With these functions, the designer is able to build solid models in a computer as he likes.

In DESIGNBASE, the features for surface processing are incorporated in the solid modeler. Generally, surface models are designed by a surface modeler and solid models by a solid modeler. By integrating these conventional functions of the two kinds of modelers, DESIGNBASE provides a new environment for 3-dimensional CAD systems. The following are the main functions of DESIGNBASE:

- **Creation of curves and surfaces**

  DESIGNBASE supports the many kinds of curves and surfaces required in any design field. In particular, DESIGNBASE introduces Gregory patches and rational boundary Gregory patches, the advantage of which is smooth connection of surfaces.

- **Local modification operations**

  This function modifies part of an object by creating, deleting, moving edges and vertices, or lifting surfaces. The target to be modified is only part of an object, and thus the operation can be executed very quickly.

- **Boolean operations**

  This function calculates the union, difference, and intersection between two objects including those with natural quadric surfaces or free-form surfaces. The objects are never approximated by polygons even when their surfaces intersect, thus the objects can be processed or analyzed with high precision.

- **Rounding operation**

  The rounding operation of DESIGNBASE can be applied not only to polyhedrons but also to objects that consist totally of curved surfaces, where fillet surfaces are created. DESIGNBASE supports four kinds of fillet surfaces, e.g., the fillet surface created by the rolling-ball method, which uses the trajectory of the user-defined ball on two surfaces.

- **UNDO · REDO/ReExecution function**

  DESIGNBASE stores the creation and modification operations of objects, so that shapes can easily be recreated. By using the design history, objects with similar shapes (such as objects with edges of different lengths) can also be easily created.

- **Shading function**

  This is a function to display shaded images of object shapes for better recognition. DESIGNBASE supports high-speed polygon shading, as well as scan line and ray tracing methods that realize texture similar to real objects. A mapping function also gives realistic images.

Plate 1 in the frontispiece shows an example of the DESIGNBASE screen, in which one of the design stages in an automobile is displayed. Plate 2 is a copy machine designed by DESIGNBASE. The pot plant was also designed as a solid model.

3.3 Software Structure of DESIGNBASE

DESIGNBASE consists of many modules[34]. Examples include a module for operating and referencing solid model data, a module for controlling the design history for the UNDO and REDO operations, and a module for controlling the screen. Figure 3.1 illustrates how the DESIGNBASE system is constructed.

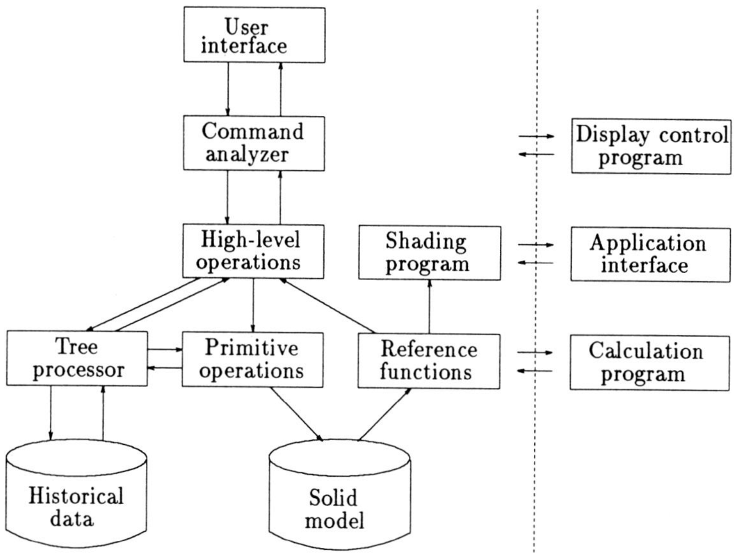

Fig. 3.1. System configuration of DESIGNBASE

The roles of these ten modules are given below:

1. **Primitive operations**: Modify shape data. Shape data can be operated only through this module.
2. **Reference functions:** Reference shape data. The shape data can be referenced only through this module.
3. **Tree processor**: Stores the primitive operations performed on the object shape and executes the UNDO·REDO/ReExecution functions. The design history of the shape data forms

the tree structure by repeating the UNDO and REDO operations. For purposes of this book the design history is called the design history tree.

4. **Calculation program**: Calculates the geometry necessary for operating shape data such as the distance between two points or the position of a point on a curve. This module includes all of calculations related to curves and surfaces and the intersection calculations between them.
5. **High-level operations**: Operate the high-level functions by using the primitive operations and reference functions. This module includes high-level local modification operations such as lifting a surface, Boolean operations, rounding operations between curved surfaces, or calculation of the volume and center of gravity of an object.
6. **Display control program**: Controls data input from the screen and data output to the screen. This module includes the display of model data, data related to the screen and mouse, or command inputs from the keyboard.
7. **Shading program**: Executes shading operations such as polygon shading, scan line shading, and ray tracing.
8. **Application interface**: Interfaces with other systems. This module writes files of shape data in the IGES format and reads data from 2-dimensional CAD systems.
9. **Command analyzer**: Analyzes input commands or arguments and calls functions suitable for their execution. All commands are analyzed and executed in this module.
10. **User interface**: Controls command menus and their hierarchical structures. This module includes menu display on the screen and input of commands and their arguments from the menus.

The functions described in the previous section can be executed, by efficiently relating them with these ten modules. The primitive operations, which form the basis of solid modelers, are explained in detail in the following sections.

3.4 Data Structure of Boundary Representations

DESIGNBASE is a B-reps based modeler. This means that DESIGNBASE can represent an object that consists of different curves and surfaces and can execute local modifications to the object. This section outlines how boundary representations have been generally processed in computers and provides an explanation of the data structure which is actually used in DESIGNBASE.

3.4.1 Solid models with boundary representations

In a B-reps based system, a model is represented by boundary faces which separates it from the outside. In other words, a solid model based on boundary representations is comprised of elements such as faces or edges that represent the boundaries between the model (inside) and its exterior (outside). The elements necessary to represent a model with boundary surfaces are as follows[113]:

- **Solid**

  Indicates one object; the unit that brings the following five elements together.

- **Shell**

  Encloses a space. A solid is a sequence of volumes enclosed with shells.

- **Face**

  Indicates the boundary of a surface. A shell is the volume enclosed with a sequence of faces.

- **Loop**

  Indicates the boundary of a face. A face always has a loop as its outer boundary. If a face contains rings, the number of loops belonging to the face is increased by the number of the rings.

- **Edge**

  This is indicated by a line where two faces intersect, that is, the curve created when two faces intersect. A loop consists of a sequence of edges.

- **Vertex**

  Indicates the corner of a solid. Several edges meet at a vertex.

These six elements indicate only how the boundaries of a model are connected to each other. They are known as the **topological elements** (see Figure 3.2).

The topological elements never change even when a model is translated, rotated, or scaled up and down. On the other hand, there are **geometric elements** which describe the shape of a model. These are explained below:

- **Surface**

  Indicates the shape of a face in 3-dimensions. It is associated with the face in the topological elements.

- **Curve**

  Indicates the shape of a line in 3-dimensions. It is associated with the edge in the topological elements.

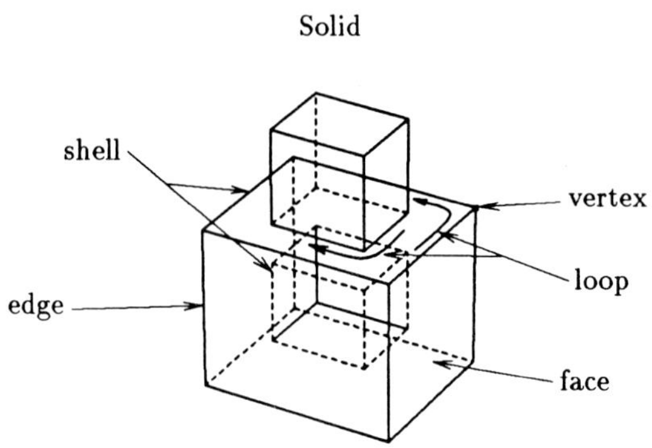

Fig. 3.2. Topological elements of a solid model in a B-reps based system

- **Point**

  Indicates coordinates in 3-dimensions. It is associated with the vertex in the topological elements.

Figure 3.3 shows two models whose topological elements are the same but geometric elements are not. As the figure shows, the shapes depend on the geometric elements.

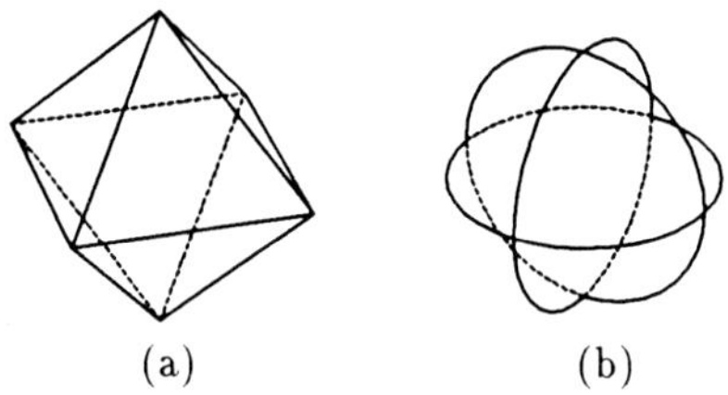

Fig. 3.3. Two solids with similar topological elements but different geometric elements

As explained so far, a solid model based on boundary representations consists of both topological and geometric elements. The relationship between these two kinds of elements is illustrated in Figure 3.4[162]. In a B-reps based system, both topological and geometric elements have to be considered. For example, when executing a Boolean operation, both the intersection curves (geometric elements) between two curved surfaces, and the edges (topological elements) created on the intersection curves have to be considered.

3.4.2 Data structures of boundary representations

This subsection describes the data structures that efficiently store B-reps based solid models. The data structure of the geometric elements of a model

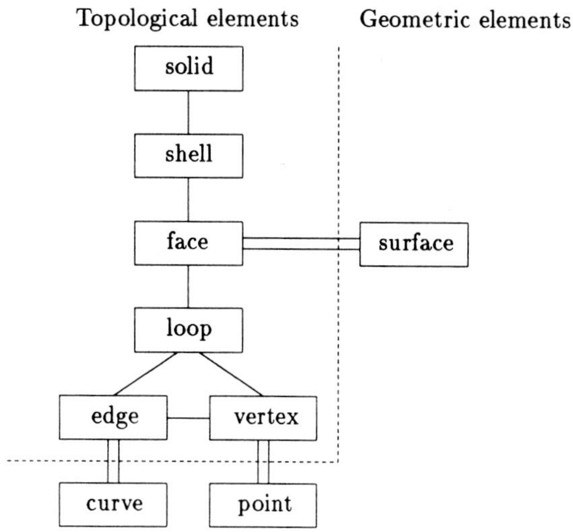

Fig. 3.4. Relationship between elements on a B-reps based solid model

can be automatically determined from the mathematical representations of the elements (for example, Bézier curve or B-spline surface). This subsection therefore focuses on the data structure of the topological elements, which specify how boundary surfaces of a model are connected.

As previously explained, the topological elements are: solid, shell, face, loop, edge, and vertex. As an example, "a solid represented by a single shell and with no rings in its faces" means a solid that has no empty space in it and for which the number of faces is equal to the number of loops. The loops represent the boundaries of the faces and in this case the two elements (face and loop) can be regarded as the same.

To represent a model completely by using three of the topological elements (face, edge, and vertex), adjacency relationships between the three elements must be quite clear, e.g., which faces are connected to either side of an edge. As Weiler presented in his paper, there are nine relationships in an adjacency topology of faces, edges, and vertices. He proved that if one of the three relationships shown in Figure 3.5 could be completely represented, then the other eight relationships would be possible[158].

(a) When a face (F) is viewed from the outside of a solid, edges (E) surrounding the face can be referenced in the same direction, usually counterclockwise (**FE adjacency relationship**).

(b) When a vertex (V) is viewed from the outside of a solid, edges (E) connected to the vertex can be referenced in the same direction, usually clockwise (**VE adjacency relationship**).

(c) When an edge (E) is viewed from the outside of a solid, other edges connected to that edge can be referenced in the same direction, usually clockwise (**EE adjacency relationship**).

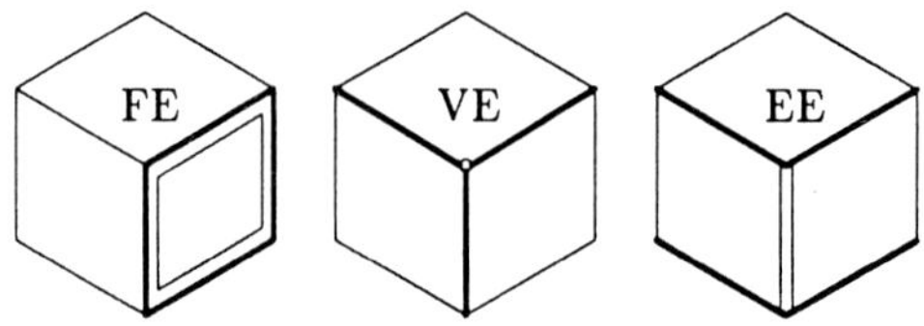

Fig. 3.5. Adjacency relationships of faces, edges, and vertices in a solid model

From the explanation included in Figure 3.5, it is evident that the data structure of any of the three adjacency relationships is sufficient when representing a solid model in a computer.

If the FE adjacency relationship shown above is represented in a computer, the topological elements may be stored as shown in Figure 3.6.

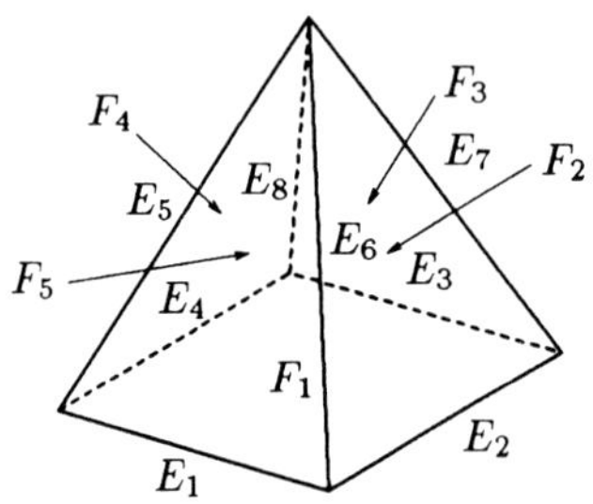

(a)

Face	Edge
F_1	E_1 - E_4 - E_3 - E_2
F_2	E_2 - E_7 - E_6
F_3	E_3 - E_8 - E_7
F_4	E_4 - E_5 - E_8
F_5	E_1 - E_6 - E_5

(b)

Fig. 3.6. Data structure in which a face owns all of its surrounding edges

In this data structure, a face owns all of its surrounding edges in such order, that the structure can be intuitively understood. However, because the number of the edges surrounding a face is not constant, the data length is variable and therefore processing in a computer can be troublesome. Furthermore, with this data structure, the algorithm to obtain the VE or EE adjacency relationships is complicated.

Baumgart[7] proposed the **winged-edge structure** to efficiently store the topological elements in boundary representations. The winged-edge structure is based on the EE adjacency relationship of a solid. The structure shown in Figure 3.7 can be uniquely defined for each edge of a solid. As this figure shows, one edge always connects to two vertices (v1, v2), two faces (f1, f2), and four **winged edges** (cwel, ccwel, cwe2, ccwe2). This means therefore that

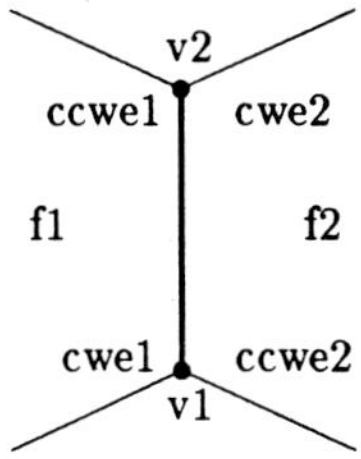

(a)

Edge Table

v1	v2
cwe1	cwe2
ccwe1	ccwe2
f1	f2

(b)

Fig. 3.7. Winged-edge structure

in the winged-edge structure, fixed-length data can be stored for each edge. This makes the winged-edge structure efficient when processing topological elements in a computer.

There is a fault, however, in the winged-edge structure. Whereas the structure is based on the EE adjacency relationship of a solid, it cannot always represent the relationships. In the case where the end vertices of an edge are the same, i.e., a self-loop exists, the nine adjacency relationships of the topological elements cannot be completely obtained. Neither can the winged-edge structure represent the state where a vertex but no edges exists in space. This is because the winged-edge structure is based on edges and thus cannot be applied to a state where there is no edge.

Weiler and Mäntylä solved these problems by proposing the data structure[90, 158] based on the separation of an element called an **edge half** from either side of an edge used in the winged-edge structure. This data structure is divided into two types: **face-edge structure** based on the FE adjacency relationship (Figures 3.8) and **vertex-edge structure** based on the VE adjacency relationship (Figure 3.9).

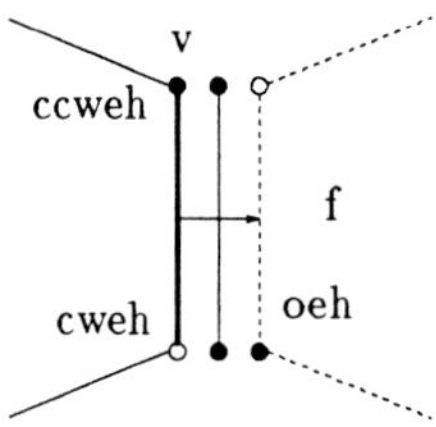

(a)

Edge Half Table

v
cweh
ccweh
f
oeh

(b)

Fig. 3.8. Face-edge structure

Mäntylä calls the face-edge structure shown in Figure 3.8 the **half-edge structure**. As shown in Figure 3.10, the face-edge structure can represent the state where only one vertex (V) exists in space.

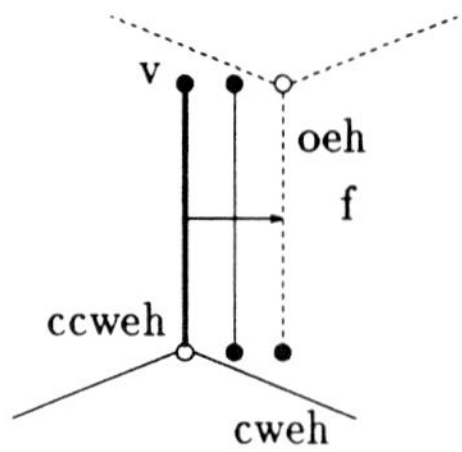

(a)

Edge Half Table

v
cweh
ccweh
f
oeh

(b)

Fig. 3.9. Vertex-edge structure

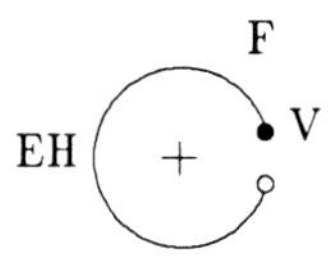

(a)

Edge Half Table

v	:	V
cweh	:	EH
ccweh	:	EH
f	:	F
oeh	:	–

(b)

Fig. 3.10. How to represent one vertex by the face-edge structure

Weiler and Woo have put forward arguments on the winged-edge and face-edge structures; about their capacities for data structures and their efficiencies to access data[158,164]. Based on these arguments, it can be concluded that the face-edge structure is superior to the winged-edge structure in memory capacity and access efficiency.

So far, four kinds of data structures in boundary representations have been explained. These data structures are based on the following two assumptions:

(i) a solid has no part which shares only edges and
(ii) a solid has no part which shares only vertices.

The **two-manifold** is a solid to which these two assumptions hold true. Usually, most of the solids designed are two-manifolds. The solids shown in Figure 3.11 are not regarded as two-manifolds because the two assumptions do not apply. Such solids may sometimes be designed especially when Boolean operations are executed. Figure (a) illustrates two prisms sharing only one edge and (b) illustrates two cones sharing only one vertex.

Solids that are not two-manifolds are called **non-manifolds**. To represent the non-manifold in Figure 3.11 (a) by the winged-edge or face-edge structure, two edges must be duplicated. Because of this requirement, the processing is very troublesome.

Weiler suggested the **radial-edge structure**[159] (an expansion of the half-edge structure), to completely represent a non-manifold. In the radial-

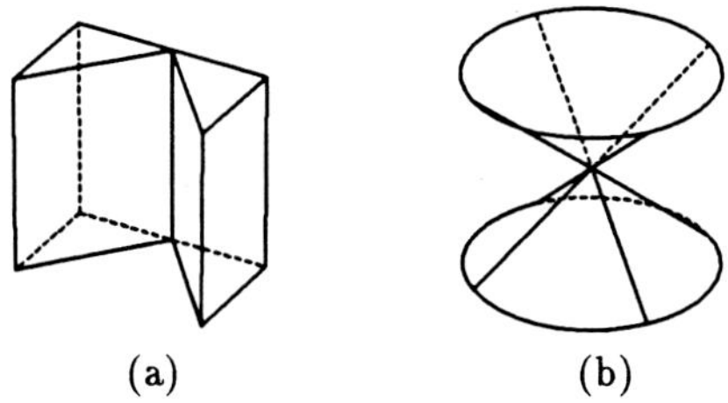

Fig. 3.11. Solids that cannot be represented as two-manifolds

edge structure, a new element **use** is created from the four topological elements: vertex, edge, face, and loop. The adjacency relationships between these four topological elements are determined by the element use. Figure 3.12 illustrates the concept of the radial-edge structure.

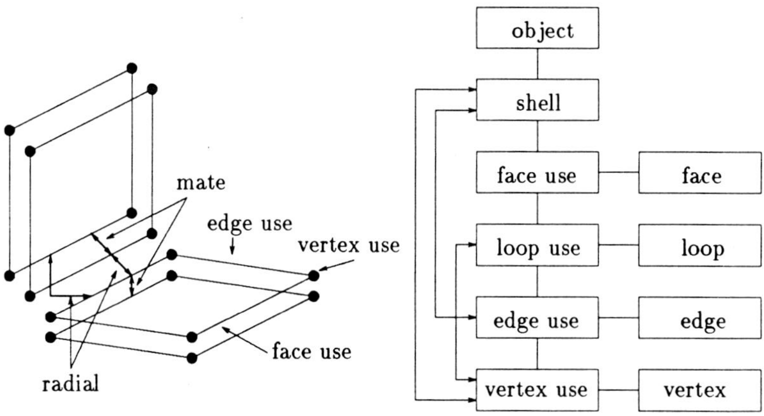

Fig. 3.12. Radial-edge structure

3.4.3 Data structure of topological elements in DESIGNBASE

In DESIGNBASE, solid model data is represented by the winged-edge structure. As shown in Figure 3.13, however, the four winged edges are not all stored in the system. The system stores only two of these edges: the end-left-edge which connects to the end point of the straight edge E, and the start-right-edge which connects to the starting point of E. This causes low efficiency in accessing data on the topological elements but less memory capacity is needed.

Furthermore, in DESIGNBASE the face and loop of the topological elements are not differentiated. They are regarded as one element, i.e., a loop. There are two types of loops: **P-loop** (parent-loop) for the outer boundaries

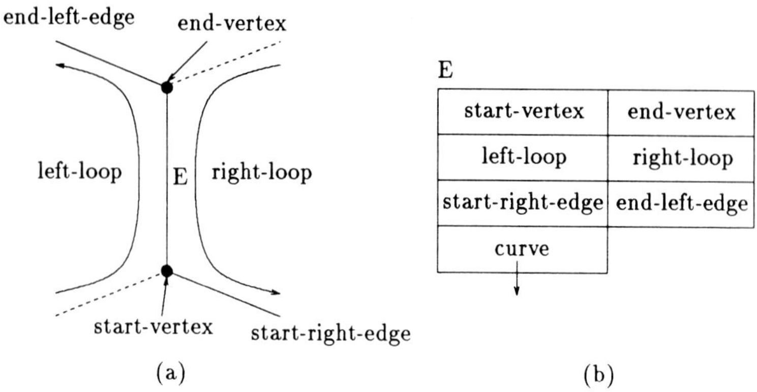

Fig. 3.13. Edge data structure in DESIGNBASE

of a face and **C-loop** (child-loop) for the inner boundaries (the rings). The data structure of a loop in DESIGNBASE is illustrated in Figure 3.14.

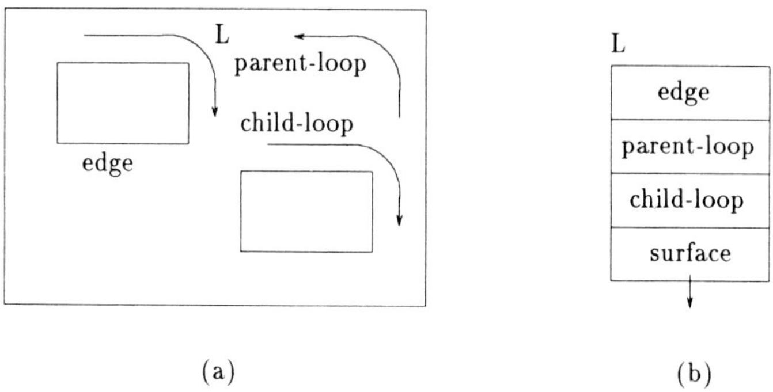

Fig. 3.14. Loop data structure in DESIGNBASE

As this data structure indicates, if a face contains three rings, the data forms the list structure shown in Figure 3.15.

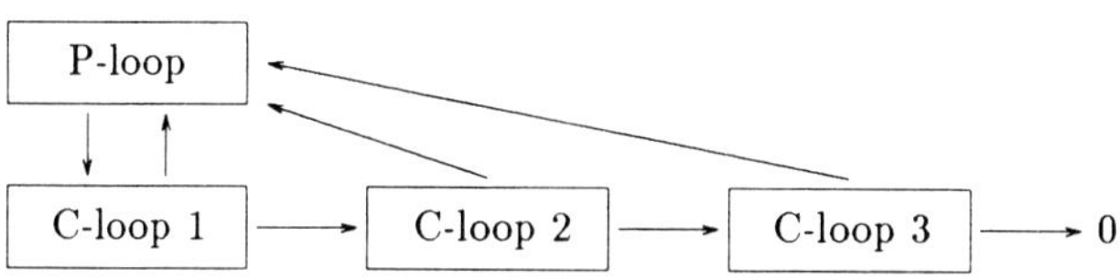

Fig. 3.15. List structure between P-loop and C-loop

The data structures of the four topological elements used in DESIGNBASE — body, loop, edge, and vertex — are shown below. In DESIGNBASE, "body" means a solid and the topological element shell is not specifically supported because a shell can be obtained from the winged-edge structure as the occasion demands.

```
struct dbBody {
    char    *name;                    /* Body name */
    int     loop_number;              /* Total number of loops */
    int     edge_number;              /* Total number of edges */
    int     vertex_number;            /* Total number of vertices */
    struct dbLoop    *loop_table;     /* Loop table */
    struct dbEdge    *edge_table;     /* Edge table */
    struct dbVertex *vertex_table;    /* Vertex table */
};

struct dbLoop {
    int     edge;                  /*
                                    * One of the edges
                                    * consisting of a loop
                                    */
    int     parent_loop;           /* P-loop */
    int     child_loop;            /* Another C-loop or 0*/
    struct dbSurface *surface;     /* Surface data of P-loop */
};

struct dbEdge {
    int     start_vertex;      /* Vertex on the start point of an edge */
    int     end_vertex;        /* Vertex on the end point of an edge */
    int     left_loop;         /* Loop on the left side of an edge */
    int     right_loop;        /* Loop on the right side of an edge */
    int     end_left_edge;     /* Left edge of the end point side */
    int     start_right_edge;/* Right edge of the start point side */
    struct dbCurve *curve;     /* Curve data of an edge */
};

struct dbVertex {
    int     edge;            /* One of the edges connected to a vertex */
    struct dbPoint point; /* Coordinates of a vertex */
};
```

3.4.4 Data structure of geometric elements in DESIGNBASE

In DESIGNBASE, geometric coordinates are stored as elements belonging to a vertex. A curve is stored as an element belonging to an edge, and a surface as an element belonging to a P-loop. Curves and surfaces are represented by type and data because they have to be processed uniformly. In other words, data on curves and surfaces can be sent and received in general form of curves and surfaces, without considering their types in the superordinate module. In this book, curves are referred to as **generalized curves**, and surfaces, as **gen-**

eralized surfaces. The following shows the data structures of coordinates, curves, and surfaces.

```
struct dbPoint {
    double  x, y, z;     /* x-, y-, and z-coordinates */
};

struct dbCurve {
    int     type;        /* Type of a curve */
    union {              /* Data of a curve */
        struct dbLine                       ln;
        struct dbArc                        ar;
        struct dbBezier_curve               bc;
        struct dbRational_Bezier_curve      rbc;
    } *data;
};

struct dbSurface {
    int type;            /* Type of a surface */
    union {              /* Data of a surface */
        struct dbPlane                      pl;
        struct dbSphere                     sp;
        struct dbCylinder                   cy;
        struct dbCone                       co;
        struct dbBezier_patch               bp;
        struct dbRational_Bezier_patch      rbp;
        struct dbGregory_patch              gp;
        struct dbRational_Gregory_patch     rgp;
    } *data;
};
```

The data regarding circular arcs and straight lines is expressed clearly in the computer according to their type. For example, a circular arc is expressed with geometric data; mainly the center and the radius. This data can be used to efficiently calculate intersection curves. Popular curves such as arcs should be stored using geometric data because this reduces processing time. Other important curves, like ellipses and parabolas, are correctly expressed by the mathematical representation which uses rational Bézier curves. If the correct mathematical form is used to represent the curves, then the intersection calculations for the curves will be correct. Since ellipses or parabolas are not often used, processing efficiency is hardly affected.

DESIGNBASE clearly stores the data of planes, spheres, cylinders, and cones as surfaces. This is because these surfaces are very popular in machine parts. They must be expressed correctly to be efficiently used in intersection calculations. DESIGNBASE prepares Bézier surfaces and rational Bézier surfaces of the nth degree for processing from external systems. For these surfaces, there is no limitation of degrees as for the internal data. Any type of surface can be processed in DESIGNBASE as long as it can be converted to a (rational) Bézier type.

One of the important requirements of surface design in CAD systems is the smooth connection of surfaces. In DESIGNBASE, this requirement is met by using the Gregory patch and rational boundary Gregory patch. These patches are very efficient when connecting a free-form surface like a Bézier surface and a quadric surface like a cylindrical surface.

When the use of many kinds of surface types is required, it can be troublesome having to process each type of curve and surface. This happens in the subordinate modules, where curves and surfaces are processed by using the optimum method for each type. To minimize this problem, generalized curves and surfaces are introduced in DESIGNBASE. For this reason also, generalized curves and surfaces are used in superordinate modules of modeling. The amount of programs necessary for curve and surface processing is also greatly reduced because the processing is unified. In this way, DESIGNBASE realizes the efficient processing of curves and surfaces.

As explained so far, DESIGNBASE supports many kinds of curve and surface types, while enabling an efficient modeling procedure and a compact system.

3.5 Primitive Operations

In DESIGNBASE, complicated modification operations such as Boolean operations are executed by a set of primitive operations. This section explains the Euler operations which are the primitive operations for solids. It also explains the primitive operations used in DESIGNBASE, which are an expansion of the Euler operations.

3.5.1 Topological elements and Euler formula

As explained earlier, the topological elements of a B-reps based solid model are solid, shell, face, loop, edge, and vertex. To represent a solid model in a B-reps based system, these six elements should be controlled so that there is no incongruence between them. However, when a model is modified by directly rewriting its topological elements, it is difficult to keep the consistency between the topological elements. Mäntylä and Wilson pointed out[91,162] that a model which is consistent must satisfy the **Euler formula**. Figure 3.16 illustrates the Euler formula.

When a face has no rings or holes i.e., the number of loops is equal to the number of faces, then

$$v - e + f = 2s$$

where v indicates the number of vertices, e the number of edges, f the number of faces,

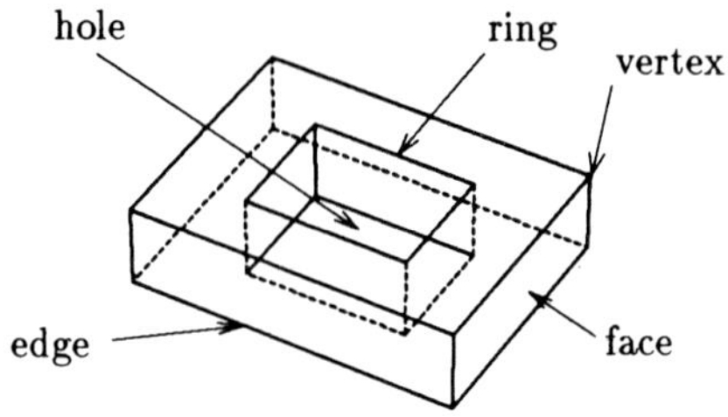

$$v - e + f - r = 2(s - h)$$
$$\rightarrow 16 - 24 + 10 - 2 = 2(1 - 1)$$

Fig. 3.16. Euler formula

When faces have rings i.e., the number of loops is greater than the number of faces, then

$$v - e + f - r = 2(s - h)$$

where r indicates the number of rings in a face, h is the number of through holes in a solid, and s, the number of shells.

To satisfy these expressions and keep the consistency between the topological elements, a complicated modification operation must be broken down into several primitive operations.

3.5.2 Euler operations

The Euler operations used to efficiently perform operations on solid models was proposed by Baumgart[8] who developed the winged-edge structure. The Euler operations are a set of primitive operations that maintain the consistency between the topological elements of a solid. According to Wilson[162], the Euler operations can be applied to wireframe models as well as solid models. This means therefore that the operations can be applied to various types of modelers other than solid modelers. At present, the Euler operations are supported in many B-reps based solid modelers. For example, Geometric Workbench (GWB) developed by Mäntylä uses Euler operations as the basis of modification operations[91]. The Euler operations can create and delete topological elements such as faces, edges, or vertices of a solid. The operators and their respective operations are shown in Table 3.1 and Figure 3.17[90].

Of the many advantages associated with the Euler operations, the two major ones are:

- The complicated processing required to maintain the consistency between the topological elements of a model can be concealed from the outside. In other words, the processing operation can be executed from the outside without directly operating the data which keeps the relationship between the topological elements constant.

Table 3.1. Euler operations

Euler Operator		Explanation	
mvfs	kvfs	make vertex face shell	kill vertex face shell
mev	kev	make edge vertex	kill edge vertex
mef	kef	make edge face	kill edge face
kemr	mekr	kill edge make ring	make edge kill ring
kfmrh	mfkrh	kill face make ring hole	make face kill ring hole

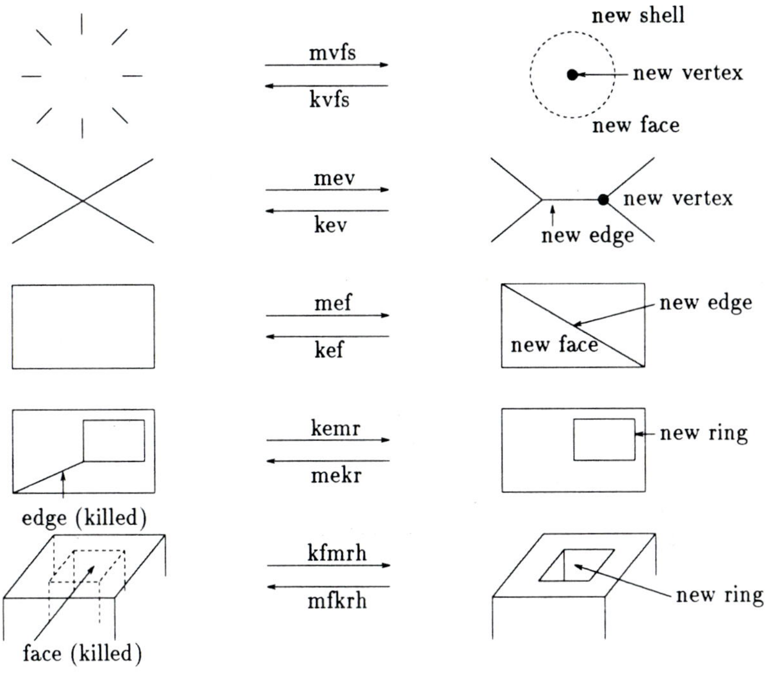

Fig. 3.17. Concept of Euler operations

- Each primitive operation has a reverse operation, and by storing the process of the operations, modifications can be easily cancelled.

As already explained, a model with correct topology satisfies the Euler formula and the Euler operations can always create a model that satisfies the Euler formula. Moreover, modification operations can be cancelled by executing their reverse operations as complicated modification operations are decomposed into sequences of primitive operations named Euler operations.

3.5.3 Primitive operations in DESIGNBASE

The **primitive operations** in DESIGNBASE consist of topological operations (based on Euler operations) and the following two operations:

(i) geometric operations such as changing an edge into a curve or moving a vertex, and

(ii) global operations such as translating a whole solid or scaling it up and down[34].

To create a model or to correct its shape in the system, these primitive operations have to be used. Because, all of the primitive operations have reverse operations, once the executed primitive operations are stored in the system, the UNDO and REDO functions can be used for modification operations.

Let us first consider the topological operations. All of the operations set out below are applied to a body B, which means a solid in DESIGNBASE.

- **MEVVL**($B\downarrow$, $E\uparrow$, $V_1\uparrow$, $V_2\uparrow$, $L\uparrow$, $\mathbf{p}_1\downarrow$, $\mathbf{p}_2\downarrow$)

  KEVVL($B\downarrow$, $E\downarrow$, $V_1\uparrow$, $V_2\uparrow$, $L\uparrow$, $\mathbf{p}_1\uparrow$, $\mathbf{p}_2\uparrow$)

  MEVVL is the operation to create a straight edge E that connects points $\mathbf{p}_1$ and $\mathbf{p}_2$ in empty space. KEVVL is its reverse operation. MEVVL creates two vertices V_1 and V_2, and a loop L, as well as an edge E (Figure 3.18).

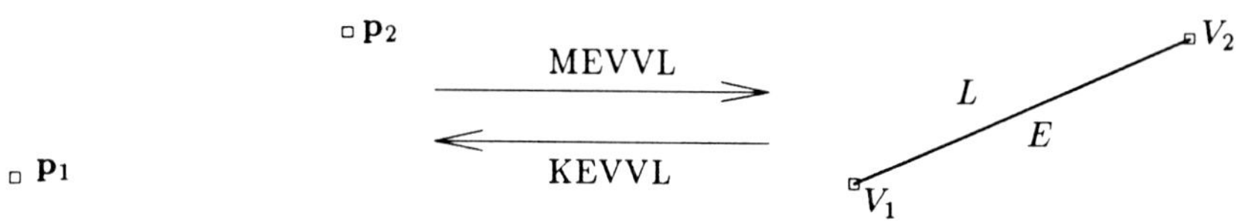

Fig. 3.18. MEVVL and KEVVL

- **MEV**($B\downarrow$, $E\uparrow$, $V_1\uparrow$, $V_2\downarrow$, $L\downarrow$, $\mathbf{p}\downarrow$)

  KEV($B\downarrow$, $E\downarrow$, $V_1\downarrow$, $V_2\uparrow$, $L\uparrow$, $\mathbf{p}\uparrow$)

  MEV is the operation to create a straight edge E from a vertex V_2 on a loop L to a point $\mathbf{p}$ in space. KEV is its reverse operation. MEV also creates a new vertex V_1 on the point $\mathbf{p}$ (see Figure 3.19). The newly-created edge E belongs to the loop L.

- **MEL**($B\downarrow$, $E\uparrow$, $L_1\uparrow$, $L_2\downarrow$, $V_1\downarrow$, $V_2\downarrow$)

  KEL($B\downarrow$, $E\downarrow$, $L_1\uparrow$, $L_2\uparrow$, $V_1\uparrow$, $V_2\uparrow$)

  MEL is the operation to create a straight edge E that connects two vertices V_1 and V_2 on a loop L_2. KEL is its reverse operation. MEL creates a new loop L_1 by splitting L_2, after creating the edge E (see Figure 3.20).

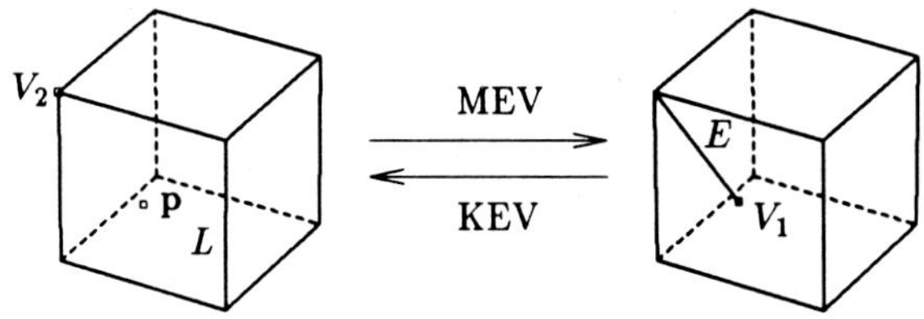

Fig. 3.19. MEV and KEV

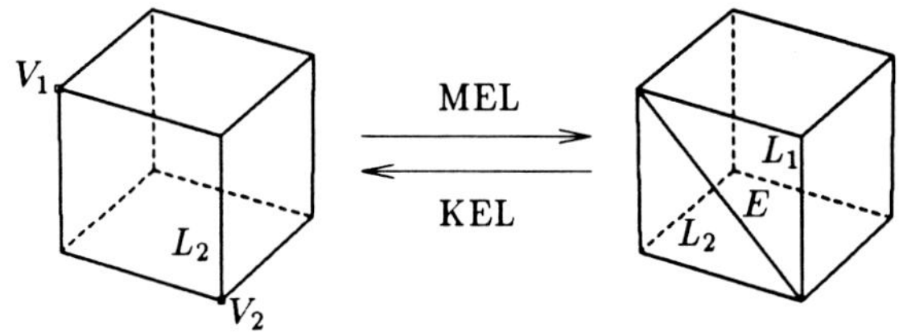

Fig. 3.20. MEL and KEL

- **MEKL**($B\downarrow$, $E\uparrow$, $L_1\downarrow$, $L_2\downarrow$, $V_1\downarrow$, $V_2\downarrow$)
  KEML($B\downarrow$, $E\downarrow$, $L_1\uparrow$, $L_2\uparrow$, $V_1\uparrow$, $V_2\uparrow$)
  MEKL is the operation to create a straight edge E that connects a vertex V_1 on a loop L_1 to a vertex V_2 on a loop L_2. KEML is its reverse operation. MEKL deletes the loop L_1, after creating the edge E (Figure 3.21). On the other hand, KEML creates a new loop L_1 by splitting L_2, after deleting the edge E. This operation is useful when creating an edge between a P-loop and a C-loop.

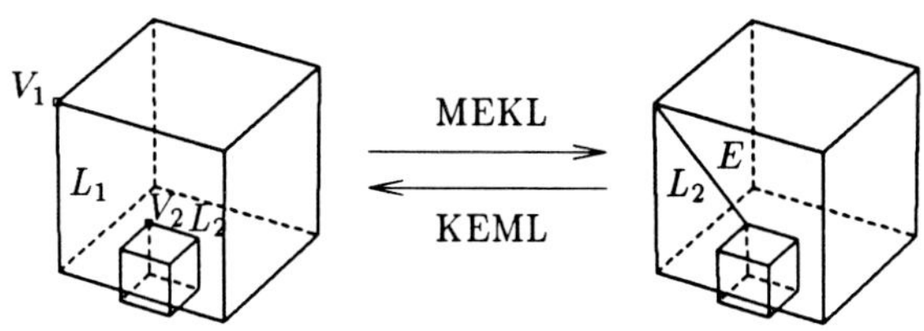

Fig. 3.21. MEKL and KEML

- **MVE**($B\downarrow$, $V\uparrow$, $E_1\uparrow$, $E_2\downarrow$, $\mathbf{p}\downarrow$)
  KVE($B\downarrow$, $V\downarrow$, $E_1\uparrow$, $E_2\uparrow$, $\mathbf{p}\uparrow$)
  MVE is the operation to create a vertex V at a point $\mathbf{p}$ on an edge E_2. KVE is its reverse operation. MVE divides the edge E_2 and creates a new edge E_1 (Figure 3.22). The edge can be of the curve type.

- **MDEV**($B\downarrow$, $E_1\uparrow$, $V_1\uparrow$, $V_2\downarrow$, $E_2\downarrow$, $E_3\downarrow$)

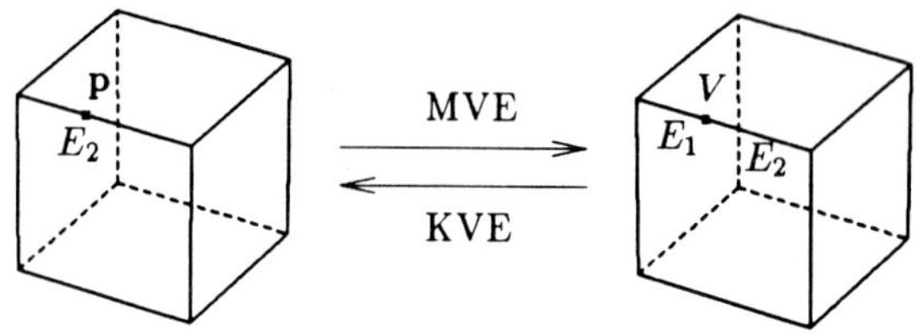

Fig. 3.22. MVE and KVE

KDEV($B\downarrow$, $E_1\downarrow$, $V_1\uparrow$, $V_2\uparrow$, $E_2\uparrow$, $E_3\uparrow$)

MDEV is the operation to create an edge of length 0 E_1 by specifying a vertex V_2 and two edges E_2 and E_3. KDEV is its reverse operation. MDEV divides the vertex V_2 and creates a new vertex V_1 and an edge E_1 (Figure 3.23). The edges E_2 and E_3 form two of the winged edges of E_1.

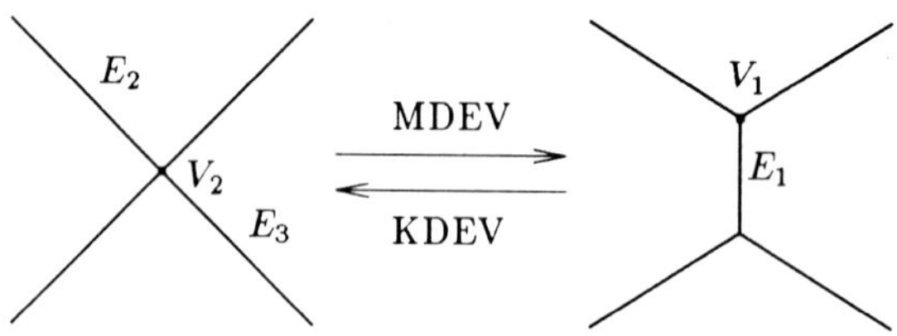

Fig. 3.23. MDEV and KDEV

- **MCLKPL**($B\downarrow$, $L_1\downarrow$, $L_2\downarrow$)

  KCLMPL($B\downarrow$, $L_1\uparrow$, $L_2\downarrow$)

  MCLKPL is the operation to convert a loop L_2 into a C-loop of the P-loop L_1. KCLMPL is its reverse operation. The loop L_2 becomes a ring of the P-loop L_1 (Figure 3.24). KCLMPL converts a loop L_2 from a C-loop to a P-loop.

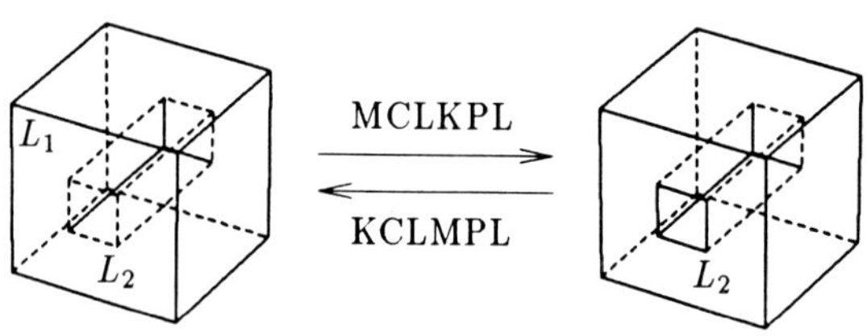

Fig. 3.24. MCLKPL and KCLMPL

Topological operations are not able to modify geometric shapes, so, DESIGNBASE supports seven geometric operations such as the operation to convert a straight line into a curve. These operations modify only geometric elements, as follows:

- **MCV**($B\downarrow$, $E\downarrow$, $c\downarrow$)
  KCV($B\downarrow$, $E\downarrow$, $c\uparrow$)
  MCV is the operation to convert a straight edge E into a curved edge containing the shape specified by a curve c. KCV is its reverse operation (Figure 3.25).

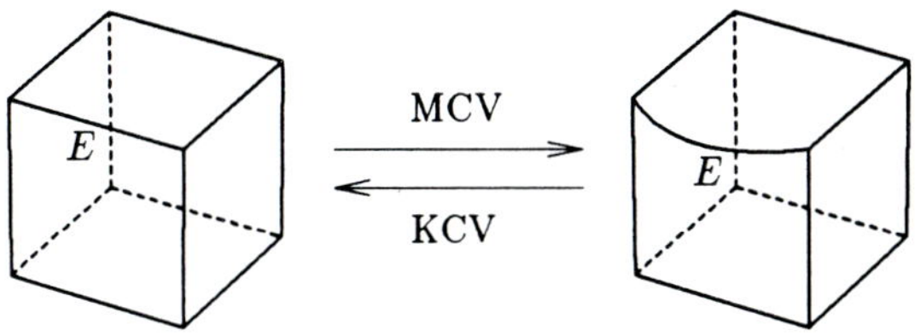

Fig. 3.25. MCV and KCV

- **MSF**($B\downarrow$, $L\downarrow$, $s\downarrow$)
  KSF($B\downarrow$, $L\downarrow$, $s\uparrow$)
  MSF is the operation to add the surface data s to a repatched P-loop L, and KSF is its reverse operation. A repatched loop is a loop whose data is not specifically held in the DESIGNBASE system and the surface of the loop is interpolated from the surrounding edges on demand (see Figure 3.26). The system stores the surface data if it is added to a loop by MSF.

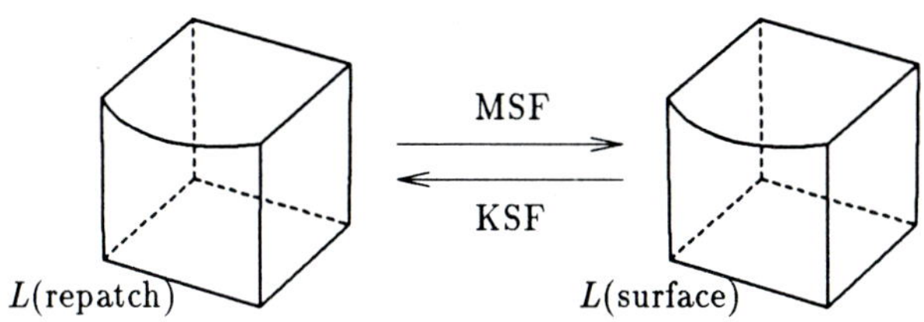

Fig. 3.26. MSF and KSF

- **TV**($B\downarrow$, $V\downarrow$, $\mathbf{v}\downarrow$)
  TV is the operation to translate a vertex V by a vector $\mathbf{v}$. The reverse operation is also TV (Figure 3.27).

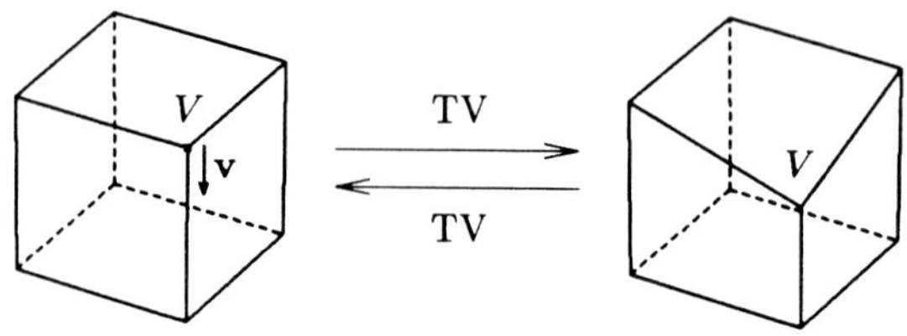

Fig. 3.27. TV

- **TCV**($B\downarrow$, $E\downarrow$, $n\downarrow$, $\mathbf{v}\downarrow$)

  TCV is the operation to translate the nth control point of a free-form edge E by a vector $\mathbf{v}$. The reverse operation is also TCV (Figure 3.28). This operation can transform the shape of a curved edge, but in doing so the data on surfaces around the edge is lost and the loops of the surfaces become repatched.

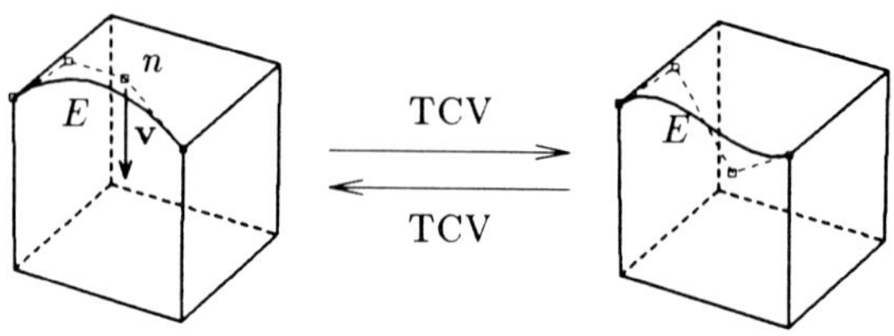

Fig. 3.28. TCV

- **TSF**($B\downarrow$, $L\downarrow$, $n\downarrow$, $\mathbf{v}\downarrow$)

  TSF is the operation to translate the nth control point of a free-form surface loop L by a vector $\mathbf{v}$. The reverse operation is also TSF. This operation can transform the shape of a surface (Figure 3.29).

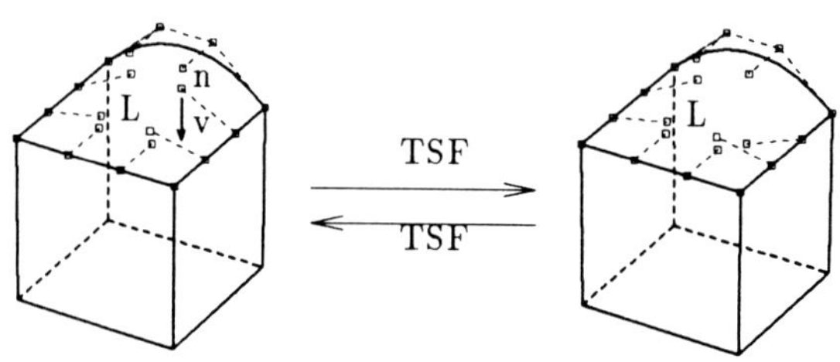

Fig. 3.29. TSF

Finally let us consider the global operations, which are performed on a body as a whole. There are seven kinds of operations as follows:

- **MOVE**($B\downarrow$, $\mathbf{v}\downarrow$)

  This is the operation to translate a body B by a vector $\mathbf{v}$. The reverse operation is also MOVE.

- **ROTATE**($B\downarrow$, $\mathbf{p}\downarrow$, $\mathbf{v}\downarrow$, $a\downarrow$)

  This is the operation to rotate a body B by the angle a about an axis that passes through a point $\mathbf{p}$ and has the direction of a vector $\mathbf{v}$. The reverse operation is also ROTATE.

- **SCALE**($B\downarrow$, $\mathbf{p}\downarrow$, $\mathbf{v}\downarrow$, $f\downarrow$)

  This is the operation to scale a body B, f times with a reference point $\mathbf{p}$ in the direction indicated by a vector $\mathbf{v}$. The reverse operation is also SCALE.

- **MIRROR**($B\downarrow$, $\mathbf{p}\downarrow$, $\mathbf{v}\downarrow$)

  This is the operation to convert a body B into a mirror image about a plane that passes through a point $\mathbf{p}$ and has a normal vector $\mathbf{v}$. The reverse operation is also MIRROR.

- **NEGATE**($B\downarrow$)

  This is the operation to turn a body B inside-out, and vice versa. For example, the operation converts a cube into a cavity of the cubic shape. The reverse operation is also NEGATE.

- **ADD**($B_1\downarrow$, $B_2\downarrow$)

  SEPARATE($B_1\downarrow$, $B_2\downarrow$)

  ADD is the operation to join a body B_2 to another body B_1. Its reverse operation is SEPARATE. These operations only change the data structures of bodies but the actual shapes never change. These operations are used in the Boolean operations.

In DESIGNBASE these primitive operations are first combined and then the local modification operations or the Boolean operations are executed. A detailed explanation of how these operations are executed is described in Chapters 7, 8, and 9.

3.6 An Example of Primitive Operations

This section describes how the primitive operations used in DESIGNBASE are constructed. The "MEVVL" and "MEV" operations are used as examples and the programs shown below are simplifications of the actual programs. Attention is paid only to the updating of the topological data. The processing for UNDO and REDO operations and that for geometric data are not explained here.

First, we look at the function "prMevvl". This function is programmed based on the MEVVL operation. This operation creates a straight line edge by using the data of two vertices, and it is the simplest of the many topological operations. Note that dbBodyList is the array of the dbBody type and the memory area for this array is allocated in advance.

```
/*
 * prMevvl --- Make an edge, two vertices, and a loop
 *
 *  Parameters:
 *    body    (I)  Body ID
 *    edge    (O)  ID of the created edge
 *    vertex1 (O)  ID of the start vertex of the created edge
 *    vertex2 (O)  ID of the end vertex of the created edge
 *    loop    (O)  ID of the created loop
 *    point1  (I)  Coordinates of the start vertex
 *    point2  (I)  Coordinates of the end vertex
 */
void
prMevvl(body, edge, vertex1, vertex2, loop, point1, point2)
    int body,*edge, *vertex1, *vertex2, *loop;
    struct dbPoint *point1, *point2;
{
    struct dbEdge *pE;
    struct dbLoop *pL;
    struct dbVertex *psV, *peV;

    /*
     * Set the edge ID to be generated to "edge".
     * Set the pointer for the memory area
     * to store the "edge" data to "pE".
     */
    EGetNewStruct(body, edge, pE);

    /*
     * Set the loop ID to be generated to "loop".
     * Set the pointer for the memory area
     * to store the "loop" data to "pL".
     */
    LGetNewStruct(body, loop, pL);

    /*
     * Set the start vertex ID to be generated to "vertex1".
     * Set the pointer for the memory area
     * to store the "vertex1" data to "psV".
     */
    VGetNewStruct(body, vertex1, psV);
```

```
    /*
     * Set the end vertex ID to be generated to "vertex2".
     * Set the pointer for the memory area
     * to store the "vertex2" data to "peV".
     */
    VGetNewStruct(body, vertex2, peV);

    /*
     * Make the edge data.
     * Define the vertex of smaller ID
     * as the start vertex of this edge.
     */
    if (*vertex1 < *vertex2) {
        pE->start_vertex = *vertex1;
        pE->end_vertex = *vertex2;
    } else {
        pE->start_vertex = *vertex2;
        pE->end_vertex = *vertex1;
    }
    pE->right_loop = pE->left_loop = *loop;
    pE->start_right_edge = pE->end_left_edge = *edge;

    /*
     * Make the loop data.
     */
    pL->edge = *edge;
    pL->parent_loop = *loop;
    pL->child_loop = 0;

    /*
     * Make the start vertex data.
     */
    psV->edge = *edge;
    psV->point = *point1;

    /*
     * Make the end vertex data.
     */
    peV->edge = *edge;
    peV->point = *point2;
}
```

The following is a fragment of the program for the function "prMev", based on the MEV operation. This function adds an edge to two vertices in a loop. This operation needs to obtain a winged edge from the straight line edge created by the function "prGtWing". The processing of the function "prMev" is complicated, so the details are omitted and only the method of updating the topological data is shown in here.

```
/* prMev --- make an edge and a vertex
 *
 *  Parameters:
 *    body      (I)  Body ID
 *    edge      (O)  ID of the created edge
 *    vertex1   (O)  ID of the created vertex
 *    vertex2   (I)  ID of the start vertex of the edge to be created
 *    loop      (I)  ID of a loop that contains the edge to be created
 *    point     (I)  Coordinates of "vertex1"
 */
void
prMev(body, edge, vertex1, vertex2, loop, point)
    int body, *edge, *vertex1, *vertex2, *loop;
    struct dbPoint *point;
{
    int *wEdge1;
    int *wEdge2;
    struct dbVertex *pnV;
    struct dbEdge *pE, *poE;

    /*
     * Set the edge ID to be generated to "edge".
     * Set the pointer for the memory area
     * to store the "edge" data to "pE".
     */
    EGetNewStruct(body, edge, pE);

    /*
     * Set the vertex ID to be generated to "vertex1".
     * Set the pointer for the memory area
     * to store the "vertex1" data to "pnV".
     */
    VGetNewStruct(body, vertex1, pnV);

    /*
     * Set the vertex data.
     */
    pnV->edge = *edge;
    pnV->point = *point;

    /*
     * Get the winged edge data of the edge to be generated
     * and set the data to wEdge2 and wEdge1;
     * obtain the IDs of the two edges
     * that surrounded a line segment * which is created
     * by connecting two vertices "vertex1" and "vertex2".
     */
    prGtWing(body, *vertex2, *vertex1, *loop, wEdge2, wEdge1);
```

```
    /*
     * Set the topological data of the edge to be generated.
     * Define the vertex of smaller ID
     * as the start vertex of this edge.
     */
    if (*vertex2 < *vertex1) {
        pE->start_vertex = *vertex2;
        pE->end_vertex = *vertex1;
        pE->start_right_edge = *wEdge1;
        pE->end_left_edge = *edge;
    } else {
        pE->start_vertex = *vertex1;
        pE->end_vertex = *vertex2;
        pE->start_right_edge = *edge;
        pE->end_left_edge = *wEdge1;
    }
    pE->right_loop = pE->left_loop = *loop;

    /*
     * Update correctly the data of the obtained winged edge.
     */
    poE = &dbBodyList[body].edge[*wEdge2];

    if (poE->start_vertex == *vertex2)
        poE->start_right_edge = *edge;
    else /* (poE->end_vertex == *vertex2) */
        poE->end_left_edge = *edge;
}
```

4. FREE-FORM CURVES

4.1 Introduction

Today's automobiles and household electrical appliances are being designed with complicated shapes and surfaces. To design such shapes and surfaces in a computer, CAD systems have been introducing various kinds of free-form curves besides straight lines, circles, ellipses, and parabolas. Representing free-form curves in a computer has become one of the most interesting areas of study in the CAD field.

Curves used in a CAD system should satisfy the following conditions:

- The shape that a designer imagines can be easily represented.
- The shape of curves can easily be modified locally or globally.
- Curves can be connected smoothly.

To satisfy these conditions, CAD systems accept various types of curves, including Bézier curves, B-spline curves, expanded forms of the rational Bézier curves, and NURBS. This chapter gives an explanation of these curves by presenting their mathematical definitions, their characteristics, and the methods used to connect each type of curve.

4.2 Parametric Curves

Representation of a curve can be accomplished by one of two methods: algebraic representation or parametric representation. This section explains these two representation methods.

4.2.1 Algebraic representation

An explicit function is one of the simplest ways of representing a curve mathematically. Let us take a planar curve as an example.

The equation given by

$$y = f(x)$$

expresses the relationship of a one-to-one correspondence, i.e. only one y is given by one x. For example, a straight line can be expressed by using the explicit function as follows:

$$y = ax + b \qquad \text{(where } a \text{ and } b \text{ are real numbers)}$$

This expression, however, cannot be applied to a closed curve such as a circle or to a curve that gives several y values corresponding to an x. Such curve is expressed as

$$f(x, y) = 0$$

This expression is called an **implicit function**. For example, in general a quadratic curve is expressed implicitly as

$$ax^2 + bxy + cy^2 + dx + ey + f = 0 \tag{4.1}$$

where a, b, c, d, e, and f are real numbers. An equation in this form can be used to express straight lines, or curves such as parabolas, hyperbolas, and ellipses. Among the curves, parabolas, hyperbolas, and ellipses are called **conic sections**. This name is derived from the cross-sectional curve of a cone which is generated by splitting the cone with a plane that does not pass through its vertex. To express a circular arc using Equation (4.1), coefficients should satisfy the condition, $a = c \neq 0$ and $b = 0$. Therefore,

$$a(x + d/2a)^2 + a(y + e/2a)^2 = (d^2 + e^2)/4a - f$$

Note that the right-hand side of the equation should be positive.

Use of the explicit and implicit function expressions really depends on the choice of coordinate systems. However, a curve cannot be divided into equal-length curves even if it is divided by the constant intervals of x or y, so it may not be drawn exactly on a screen. Especially, in implicit representations, where a one-to-one correspondence cannot always be given between x and y values, a variable cannot uniquely define the position of a curve. This problem is solved in parametric representation.

4.2.2 Parametric representation

A curve is parameterized if each of the x and y coordinates of a point on the curve is represented by a parameter. For example, each of the two coordinates x and y of a point on a planar curve is expressed by a parameter t as

$$x = x(t), \; y = y(t)$$

and the position vector of this point is expressed as

$$\mathbf{P}(t) = (x(t), y(t))$$

The implicit function expression can be obtained by deleting the parameter t from this expression to obtain another relational expression of x and y.

With algebraic representation there are some problems: mathematical expressions of curves greatly change as the coordinates are transformed or else representation of the curves becomes troublesome. However, if a curve is parameterized, the mathematical expression of the curve can be easily calculated even after rotating or translating it. Moreover, a point on a closed curved can be uniquely defined from a parameter, so that the curve is easily displayed on a screen.

The shape of a parametric curve can be controlled by using the control points of the curve. The points indicated in Figure 4.1 are called the control points, and the sequence of line segments (broken lines) defined by the control points is called the **control polygon**.

A curve can be formed by smoothing its control polygon. In other words, by specifying a control polygon, a designer can easily design curves which he imagines and their shapes can also be easily modified. Compared with the algebraic representation, the parametric representation is more advantageous. This is why parametric curves are mostly used in CAD systems.

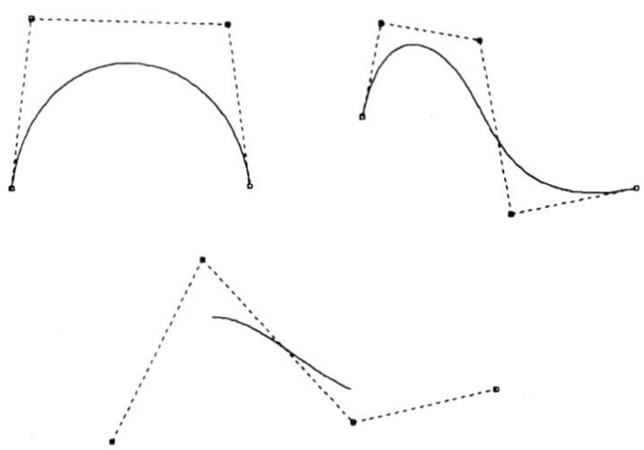

Fig. 4.1. Curves and control points

4.3 Bézier Curves

One of the most popular parametric curves is the **Bézier curve**, famous for being used to design automobiles at Rénault in France. The Bézier curve was investigated by de Casteljau[49,48] and Bézier[10] separately. The curve is named after Bézier because the thesis of de Casteljau was not available to the public. The Bézier curve can be shaped uniquely if a control polygon is defined. This makes it easy for designers to imagine the curve shape once a control polygon exists. This is why the Bézier curve is said to be designer-friendly.

4.3.1 How to represent Bézier curves

Bézier curves are formed according to the control points. The points in Figure 4.2 are the control points. In this figure, (a) illustrates a quadratic Bézier curve, (b) a quartic Bézier curve, and (c) a cubic Bézier curve.

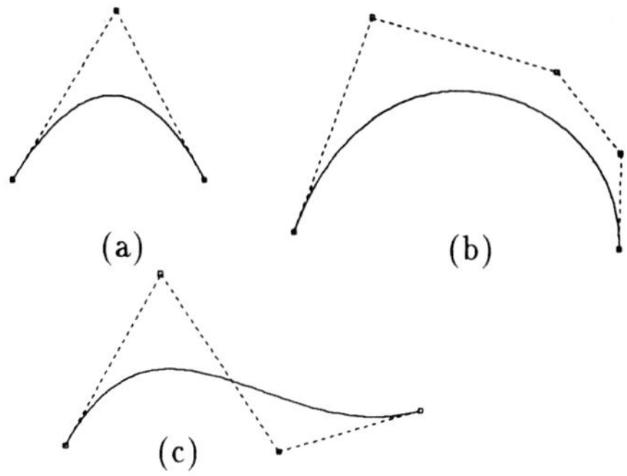

Fig. 4.2. Bézier curves

The Bézier curve $\mathbf{R}(t)$ of degree n is created from $n+1$ control points and expressed as follows:

$$\mathbf{R}(t) = \sum_{i=0}^{n} B_i^n(t)\mathbf{P}_i \qquad (0 \leq t \leq 1) \tag{4.2}$$

where $B_i^n(t)$ is the function called the **Bernstein basis**, given by

$$B_i^n(t) = \binom{n}{i} t^i (1-t)^{n-i} \tag{4.3}$$

and

$$\binom{n}{i} = \frac{n!}{(n-i)!i!}$$

indicates the binomial coefficient. Therefore, according to Equation (4.2), a point on a Bézier curve is represented as the union of control points $\mathbf{P}_i$ to which weights of the binomial coefficient are given. Figure 4.3 shows the graph of the Bernstein basis with $n = 3$.

Bézier curves have the following properties:

1. **Convex hull property**

   According to the characteristics of the Bernstein basis, the union of its values on all control points becomes 1 for any parameter value t. Equation 4.2 then becomes:

$$\sum_{i=0}^{n} B_i^n(t) = 1 \qquad (0 \leq t \leq 1)$$

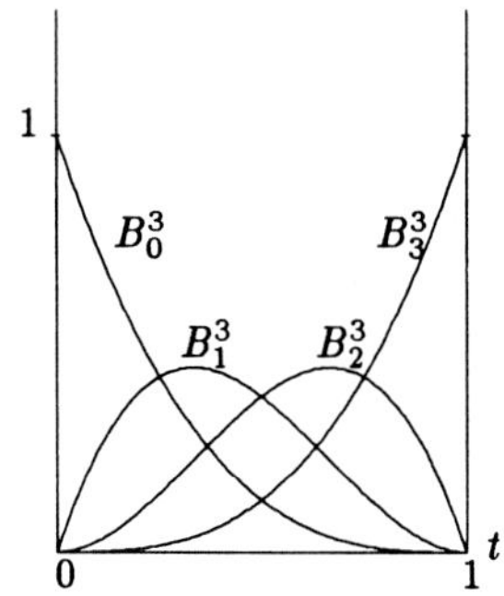

Fig. 4.3. Bernstein basis

Furthermore, the Bernstein basis is positive for an arbitrary i, such that:

$$B_i^n(t) \geq 0 \qquad (0 \leq t \leq 1)$$

Because of these characteristics, a Bézier curve is completely contained in the **convex hull** defined by the curve's control points. This condition is referred to as the **convex hull property**. The convex hull is the minimum convex area that contains all of the control points, as shown by the hatched area in Figure 4.4. The broken lines indicate the control polygons and the solid lines are the Bézier curves created by their control points.

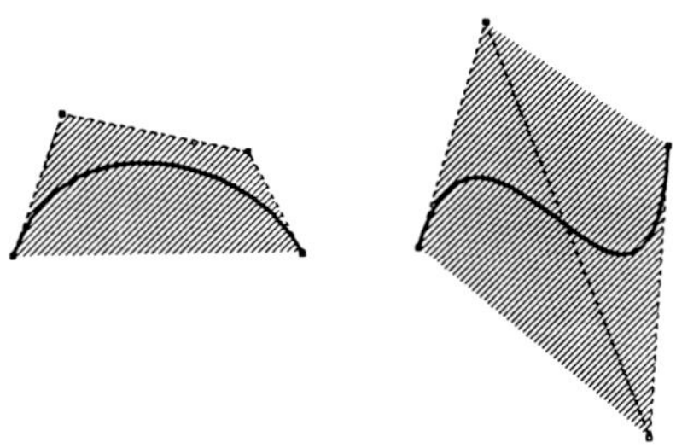

Fig. 4.4. Convex hulls of Bézier curves

In CG or CAD, a check is often made on the intersection of two models. Here, it would be time-consuming to calculate the intersection points of two free-form curves. So, if instead of the free-form curves, the convex hulls are used to check the intersection, it is possible to efficiently find curves that never intersect each other, and no time is wasted with intersection calculations.

2. **Variation diminishing property**

The shape of a Bézier curve is similar to that of its control polygon, which is formed by connecting the curve's control points with straight lines (see Figure 4.2). The Bézier curve passes through the start and end points of the control polygon and its shape is smoother than its control polygon. If an arbitrary control point is translated, the shape of the control polygon changes and thus the shape of the curve changes. The first-order derivative of the Bézier curve of degree n is obtained from Equation (4.2) as

$$\frac{d\mathbf{R}(t)}{dt} = n \sum_{i=0}^{n-1} B_i^{n-1}(t)\mathbf{a}_i \qquad (0 \leq t \leq 1) \tag{4.4}$$

where $\mathbf{a}_i$ is defined as

$$\mathbf{a}_i = \mathbf{P}_{i+1} - \mathbf{P}_i \qquad (i = 0, \cdots, n-1)$$

According to Equation (4.4), the first-order derivative at the end points of the Bézier curve is given by

$$\frac{d\mathbf{R}(0)}{dt} = n(\mathbf{P}_1 - \mathbf{P}_0) \tag{4.5}$$

and

$$\frac{d\mathbf{R}(1)}{dt} = n(\mathbf{P}_n - \mathbf{P}_{n-1})$$

These equations prove that the first-order derivative at the end points is equal to n times the size of the vector between the end point and its adjacent control point (see Figure 4.5). From this condition, it is evident that the curve shape must be largely affected by the control polygon shape.

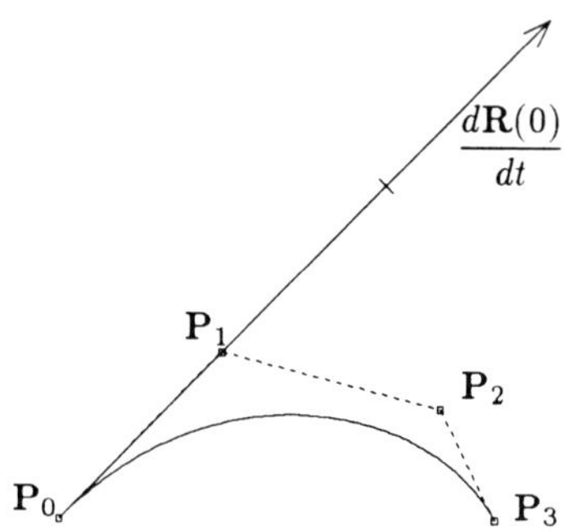

Fig. 4.5. The first-order derivative at the end point of a Bézier curve

The second-order derivative is obtained as

$$\frac{d^2\mathbf{R}(t)}{dt^2} = n(n-1)\sum_{i=0}^{n-2} B_i^{n-2}(t)\mathbf{b}_i \qquad (0 \le t \le 1) \tag{4.6}$$

where

$$\mathbf{b}_i = \mathbf{a}_{i+1} - \mathbf{a}_i \qquad (i = 0, \cdots, n-2).$$

The mth order derivative of a Bézier curve can be expressed in Bézier format by using the Bernstein basis $B_i^{n-m}(t)$.

4.3.2 Division and connection of Bézier curves

Some operations of a CAD system separate parts of a model or split a model with a plane. To effect these operations therefore, curves also need to be divided or split. The Bézier curve has the characteristics expressed by the following equation:

$$\mathbf{P}_i^r(t) = (1-t)\mathbf{P}_i^{r-1}(t) + t\mathbf{P}_{i+1}^{r-1}(t) \tag{4.7}$$

where r and i change as

$$r=1,\cdots,n$$
$$i=0,\cdots,n-r$$

and $\mathbf{P}_i^0(t) = \mathbf{P}_i$ indicates the control points of the original curve. The position vector $\mathbf{P}_0^n(t)$ defined by Equation (4.7) is the point specified by the parameter t of the Bézier curve. (This algorithm was founded by de Casteljau[48] and is known as the **de Casteljau algorithm**.)

Let us take a simple example of a cubic Bézier curve. Equation (4.7) is expanded as follows:

$$\begin{aligned}
\mathbf{P}_0^1(t) &= (1-t)\mathbf{P}_0^0(t) + t\mathbf{P}_1^0(t) \\
\mathbf{P}_1^1(t) &= (1-t)\mathbf{P}_1^0(t) + t\mathbf{P}_2^0(t) \\
\mathbf{P}_2^1(t) &= (1-t)\mathbf{P}_2^0(t) + t\mathbf{P}_3^0(t) \\
\mathbf{P}_0^2(t) &= (1-t)\mathbf{P}_0^1(t) + t\mathbf{P}_1^1(t) \\
\mathbf{P}_1^2(t) &= (1-t)\mathbf{P}_1^1(t) + t\mathbf{P}_2^1(t) \\
\mathbf{P}_0^3(t) &= (1-t)\mathbf{P}_0^2(t) + t\mathbf{P}_1^2(t)
\end{aligned} \tag{4.8}$$

The last equation $\mathbf{P}_0^3(t)$ indicates the point at the parameter t on the Bézier curve. These equations are illustated in Figure 4.6.

As this figure shows, when a curve is split at the parameter t, the control points of the two cubic Bézier curves which are generated are

$$\mathbf{P}_0^0(t), \mathbf{P}_0^1(t), \mathbf{P}_0^2(t), \mathbf{P}_0^3(t)$$

and

$$\mathbf{P}_3^0(t), \mathbf{P}_2^1(t), \mathbf{P}_1^2(t), \mathbf{P}_0^3(t)$$

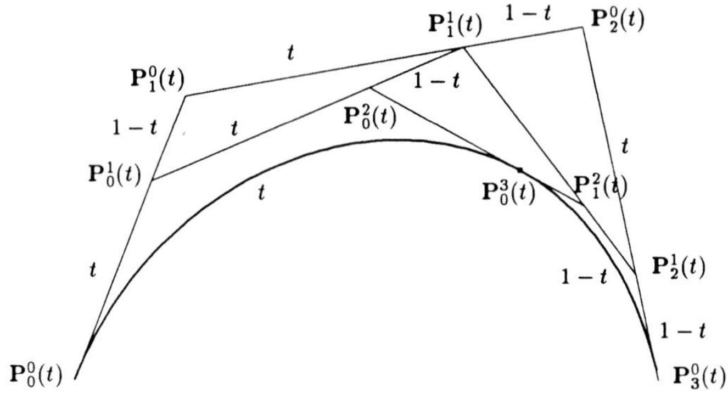

Fig. 4.6. de Casteljau algorithm

The control points of the two curves generated by the splitting can be obtained easily by repeating linear interpolations. The same can be applied to any Bézier curve of degree n. Hence, the Bézier curve is easily divided. The reverse operation can connect two Bézier curves, but because of the position of each control polygon, the two Bézier curves are not necessarily connected with a continuous curvature.

Let us now consider the continuity of two Bézier curves of the same degree whose end points coincide. Since the end points are coincident, the curves are connected with $\mathbf{C}^0$ **continuity**. If the derivative vector between the coincident end point and its adjacent control point on one curve is equal to that of the other curve, then according to Equation (4.5), the two curves are connected with $\mathbf{C}^1$ **continuity**. The two vectors may be sometimes different in length though they have the same direction. Such continuity is called $\mathbf{G}^1$ **continuity**.

4.3.3 Degree elevation of Bézier curves

The number of control points of a Bézier curve depends on its degrees. This means that to increase the curve's flexibility it is effective to elevate the degree and to increase the number of control points. Let us now consider how to elevate the degree of a Bézier curve.

Suppose a Bézier curve is expressed as

$$\mathbf{R}(t) = \sum_{i=0}^{n} B_i^n(t)\mathbf{P}_i$$

If the degree of this curve is elevated by one without changing the curve shape, the newly generated curve becomes

$$\mathbf{R}(t) = \sum_{i=0}^{n+1} B_i^{n+1}(t)\mathbf{Q}_i$$

Then, what is the relationship between the control points of the original curve $\mathbf{P}_i$ and those of the new curve $\mathbf{Q}_i$? From of the above two equations, the following equation can be obtained:

$$\sum_{i=0}^{n} \binom{n}{i} t^i (1-t)^{n-i} \mathbf{P}_i = \sum_{i=0}^{n+1} \binom{n+1}{i} t^i (1-t)^{n+1-i} \mathbf{Q}_i$$

Then, multiplying the left-hand side of this equation by $((1-t)+t)$, the equation is expanded to

$$\sum_{i=0}^{n} \binom{n}{i} (t^i (1-t)^{n+1-i} + t^{i+1} (1-t)^{n-i}) \mathbf{P}_i = \sum_{i=0}^{n+1} \binom{n+1}{i} t^i (1-t)^{n+1-i} \mathbf{Q}_i$$

By comparing the coefficients of $t^i(1-t)^{n+1-i}$ on both sides of this equation, then

$$\mathbf{Q}_i \binom{n+1}{i} = \mathbf{P}_i \binom{n}{i} + \mathbf{P}_{i-1} \binom{n}{i-1}$$

is obtained. Thus,

$$\mathbf{Q}_i = \frac{i}{n+1} \mathbf{P}_{i-1} + \left(1 - \frac{i}{n+1}\right) \mathbf{P}_i$$

The control points can be obtained from this equation when the degree of the curve is elevated by one.

4.4 Rational Bézier Curves

The Bézier curve described in the former section is the polynomial curve of degree n. Unfortunately, the precision of a curve represented by the polynomial curve is limited. For example, an arc or an ellipse cannot be precisely represented by the polynomial curve. It is also known that if an arc with radius 1 and central angle 45° is approximated by a cubic Bézier curve, the maximum distance between the arc and the curve will be about 2×10^{-6}[51]. To eliminate such short comings, the Bézier curve has been expanded into the **rational Bézier curve**.

4.4.1 How to represent rational Bézier curves

The rational Bézier curve is expressed as

$$\mathbf{R}(t) = \frac{\sum_{i=0}^{n} B_i^n(t) w_i \mathbf{P}_i}{\sum_{i=0}^{n} B_i^n(t) w_i} \qquad (0 \le t \le 1) \tag{4.9}$$

where $B_i^n(t)$ indicates the Bernstein basis expressed by Equation (4.3), $\mathbf{P}_i$ indicates the position coordinates of a control point, and w_i indicates the weight of $\mathbf{P}_i$. The relationship between the control polygon and the curve shape of the rational Bézier curve is the same as that with the Bézier curve, except for one condition: the shape of the rational Bézier curve can be changed by operating its weights. The effect of weight on the shape of the rational Bézier curve is considered in the following subsection.

4.4.2 Weights and their effects

Weight is a unique element for rational curves or surfaces. The weight of a control point is the parameter that indicates how much the curve will stretch to the control point. The larger the weight of a control point, the more the curve is stretched to this point, and vice versa. As Figure 4.7 shows, the control point of weight 0 of a curve is ignored and the infinite weight duplicates the curve at its control point and within its control polygon. If the weight is negative, the curve bounds back away from the control point. If the weights of all the control points are 1, the rational Bézier curve becomes a polynomial Bézier curve.

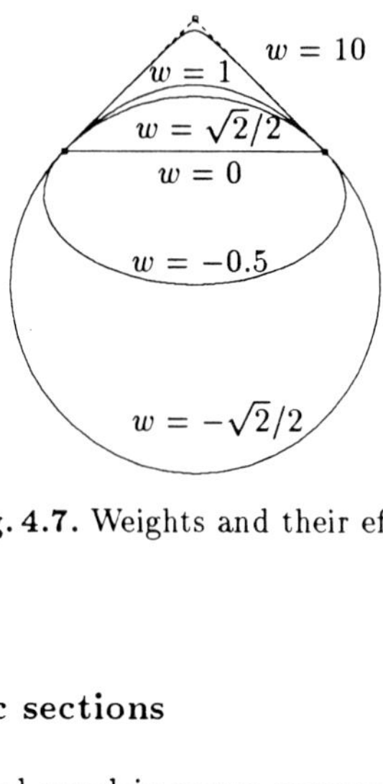

Fig. 4.7. Weights and their effects

4.4.3 Creation of conic sections

A conic section is a natural quadric curve expressed by Equation (4.1). Arcs or ellipses of the conic section are indispensable to the design of machine parts and it is therefore important in CAD systems to represent the conic sections correctly with free-form curves. This subsection explains how to represent a conic section by a rational Bézier curve.

As already known, that the rational Bézier curve that represents a conic section is quadric. Let us suppose that the three control points of the rational Bézier curve are $\mathbf{P}_0$, $\mathbf{P}_1$, and $\mathbf{P}_2$ and the weights of the control points are w_0, w_1, and w_2. Here, we define the weight w as

$$w = \frac{w_1}{\sqrt{w_0 w_2}} \tag{4.10}$$

Then, Equation (4.9) is expanded as follows[53]:

$$\mathbf{R}(t) = \frac{B_0^2(t)\mathbf{P}_0 + B_1^2(t)w\mathbf{P}_1 + B_2^2(t)\mathbf{P}_2}{B_0^2(t) + B_1^2(t)w + B_2^2(t)} \tag{4.11}$$

Furthermore, as w changes, the curve shape changes as follows[83]:

$w = 0$	Straight line
$0 < w < 1$	Elliptic curve
$w = 1$	Parabola
$w > 1$	Hyperbola

let us suppose $\mathbf{R}(t)$=$\mathbf{S}$, when $t = 1/2$. Then according to Equation (4.11), the following equation can be obtained:

$$\mathbf{S} = \frac{1}{1+w}\frac{\mathbf{P}_0 + \mathbf{P}_2}{2} + \frac{w}{1+w}\mathbf{P}_1$$

If the midpoint between $\mathbf{P}_0$ and $\mathbf{P}_2$ is defined as $\mathbf{M}$, the above equation can be arranged as

$$\mathbf{S} = (1-s)\mathbf{M} + s\mathbf{P}_1$$

where s is defined as

$$s = \frac{w}{1+w}$$

The above equation proves that the midpoint $\mathbf{S}$ for the parameter t on a quadratic curve lies on the straight line connecting $\mathbf{P}_1$ and the midpoint $\mathbf{M}$ between $\mathbf{P}_0$ and $\mathbf{P}_2$. If the position of $\mathbf{S}$ is clear, the weight w of the control point $\mathbf{P}_1$ can be obtained from

$$w = \frac{|\mathbf{M} - \mathbf{S}|}{|\mathbf{S} - \mathbf{P}_1|}$$

Let us obtain the weight of a control point of an arc whose control polygon is an isosceles triangle (see Figure 4.8). The above equation can be arranged into

$$w = \frac{|\mathbf{M} - \mathbf{S}|}{|\mathbf{P}_1 - \mathbf{M}| - |\mathbf{M} - \mathbf{S}|}$$

The tangent at $\mathbf{S}$ lies parallel to the vector $\mathbf{P}_0 - \mathbf{P}_2$, thus, if the base angle of the isosceles triangle is defined as θ, then

$$\angle \mathbf{S}\mathbf{P}_0\mathbf{M} = \frac{\theta}{2}$$

and this can be substituted in the equation to give

$$|\mathbf{M} - \mathbf{S}| = |\mathbf{P}_0 - \mathbf{M}| \tan\frac{\theta}{2} = |\mathbf{P}_1 - \mathbf{P}_0| \cos\theta \tan\frac{\theta}{2}$$

and

$$|\mathbf{P}_1 - \mathbf{M}| = |\mathbf{P}_1 - \mathbf{P}_0| \sin\theta$$

simplifying, we get

$$w = \cos\theta$$

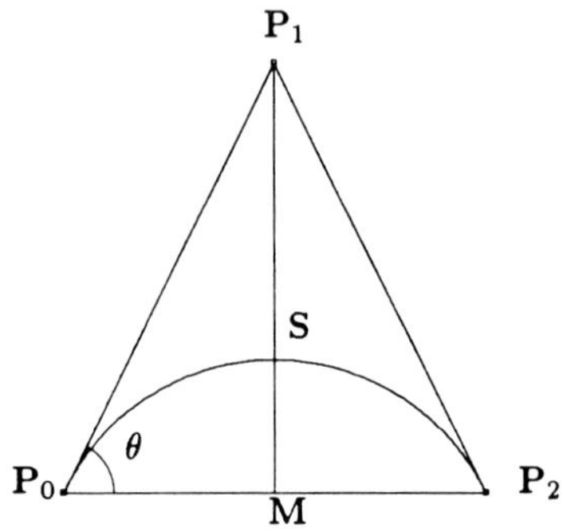

Fig. 4.8. Curve that represents an arc

4.5 B-Spline Curves

The movement of a control point affects the whole shape of a Bézier curve, thus the Bézier curve cannot be modified locally. The degrees of a Bézier curve also depends on the number of the control points. These conditions of the Bézier curve are brought about by the Bernstein basis.

The curve that compensates for these limitations is the **B-spline curve**. Schoenberg was the first to suggest the concept of a B-spline for data interpolation[133]. To make the B-spline theorem of Schoenberg more applicable, de Boor[47] and Cox[46] separately rewrote the **spline basis** into a recursive expression. Then, Riesenfeld[59,116] applied the expression to the basis of curve representation and created the B-spline curve as a new form of curve representation. Although the number of control points of a Bézier curve is automatically determined by the degrees of the curve, a B-spline curve can have any number of control points as long as this number is at least one greater than the number of degrees (degree + 1). The B-spline curve can also represent two or more segments with one curve, while the Bézier curve can represent only a polynomial curve, that is, one segment. The B-spline curve is also superior to the Bézier curve with respect to connecting curves because derivative vectors of (degree − 1)th order of the segments are equal.

4.5.1 How to represent B-spline curves

A B-spline curve is expressed as

$$\mathbf{R}(t) = \sum_{i=0}^{n} N_{i,k}(t)\mathbf{P}_i \qquad (2 \leq k \leq n+1) \tag{4.12}$$

where $N_{i,k}(t)$ is referred to as the **B-spline basis** and is defined as follows:

$$
\begin{aligned}
N_{i,1}(t) &= \begin{cases} 1 & \text{if } x_i \le t < x_{i+1} \\ 0 & \text{otherwise} \end{cases} \\
N_{i,k}(t) &= \frac{(t - x_i)N_{i,k-1}(t)}{x_{i+k-1} - x_i} + \frac{(x_{i+k} - t)N_{i+1,k-1}(t)}{x_{i+k} - x_{i+1}}
\end{aligned} \tag{4.13}
$$

In these equations, k indicates the **order** and $k-1$ indicates the **degree.** $\mathbf{P}_i$ indicates the control point and $n+1$ the number of control points. x_i indicates the **knot** and

$$x_i \le x_{i+1}$$

should be satisfied (monotonic increase). The sequence of knots

$$[x_0 \ x_1 \ \dots \ x_m]$$

is called the **knot vector.** $m+1$ indicates the number of knot vectors and the order k and the number of control points $n+1$ have the relationship

$$k + (n+1) = m + 1$$

The B-spline curve has the following characteristics:

1. **Polynomial curve**

   The B-spline curve is a polynomial curve of $k-1$ degrees between knots such that $x_i \le t < x_{i+1}$.

2. **C^{k-2} continuity**

   According to Equation (4.13), every segment of a B-spline curve is connected with C^{k-2} continuity unless the knot values are multiple knot values. The continuity of the knots whose multiplicity is p would be C^{k-p-2}.

3. **Convex hull property**

   Like the Bézier curve, the B-spline curve has the convex hull property because of the characteristic of the B-spline basis which states:

$$\sum_{i=0}^{n} N_{i,k}(t) = 1$$

$$N_{i,k}(t) \ge 0$$

4. **Variation diminishing property**

   As explained earlier, the Bézier curve follows the shape of its control polygon. The same applies to the B-spline curve and the variation of the curve is less than that of the control polygon. Moving a control point changes the curve shape by units of the **curve segment** (a curve between knots). The number of curve segments whose shape is to be changed corresponds to the order about the control point; outside of this no other curve segments can be changed. The derivation of the curve yields a recursive expression like the B-spline basis expression. For example, the first-order derivation is expressed as follows:

$$\frac{d\mathbf{R}(t)}{dt} = \sum_{i=0}^{n} N'_{i,k}(t)\mathbf{P}_i \tag{4.14}$$

$$\begin{aligned} N'_{i,1}(t) &= 0 \\ N'_{i,k}(t) &= \frac{N_{i,k-1}(t) + (t - x_i)N'_{i,k-1}(t)}{x_{i+k-1} - x_i} \\ &+ \frac{(x_{i+k} - t)N'_{i+1,k-1}(t) - N_{i+1,k-1}(t)}{x_{i+k} - x_{i+1}} \end{aligned} \tag{4.15}$$

4.5.2 Knot vectors and curves

As is evident from the curve defining expression, the basis for the coefficient of each control point is closely related to the knot vector. In other words, the curve shape is closely related to the knot vector.

1. Open uniform knot vector

   An open uniform knot vector has a multiplicity of knot values at the ends equal to the order k. Some examples are:

   - [0 0 0 1 2 3 3 3] ($k = 3$)
   - [0 0 0 0 1 1 1 1] ($k = 4$)

   The end points of a curve defined by open uniform knot vectors correspond with those of the control polygon. If the number of knots are twice as k and there are no internal knot values, (as shown in the lower line of the above example,) the B-spline curve coincides with the Bézier curve (Figure 4.9).

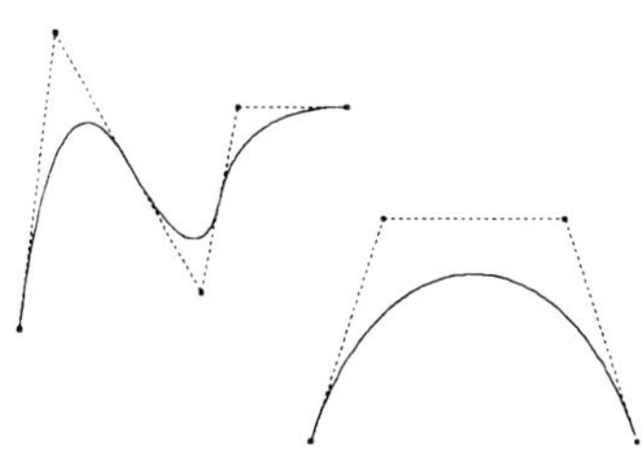

Fig. 4.9. Curves defined by open uniform knot vectors

2. Uniform knot vector

   A uniform knot vector has equally-spaced knot values. Some examples are:

- [0 1 2 3 4 5 6 7] ($k = 3$)
- [-3 -1 1 3 5 7 9 11] ($k = 4$)

The end points of a curve defined by uniform knot vectors do not correspond with those of the control polygon (Figure 4.10).

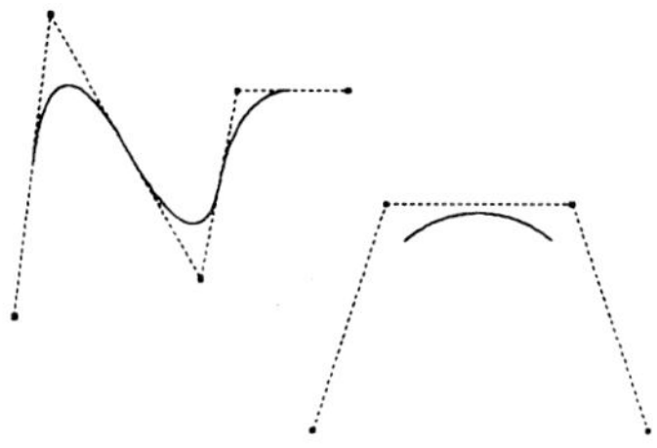

Fig. 4.10. Curves defined by uniform knot vectors

3. Nonuniform knot vector

   A nonuniform knot vector does not have a fixed format. The knot vectors that cannot be classified into either of the above two kinds are included in this format. Some examples are:

   - [0 1 2 3 3 4 5 6] ($k = 3$)
   - [0 1 2 3 4 4 4 4] ($k = 4$)

   If the multiplicity of internal knots is the same as its order (as indicated in the upper line of the above example), a **cusp**, or sharp corner, occurs at the corresponding control point, where C^0 continuity is found (Figure 4.11).

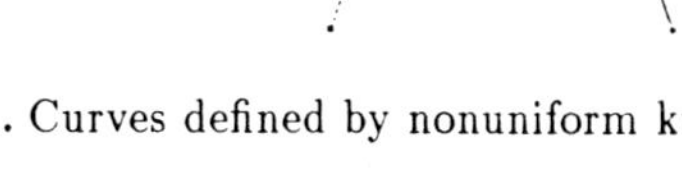

Fig. 4.11. Curves defined by nonuniform knot vectors

4.5.3 Connection and division of curves

As explained in the previous subsection, multiple internal knot values of a nonuniform knot vector cause a cusp at the corresponding control point. Accordingly, if a control point where a cusp occurs and the two adjacent control points lie on a straight line, the curves of the corresponding control polygon spans are connected smoothly. On the other hand, if a curve is to be divided, a control point should be created at the parameter to be divided so as to have multiple knot values, without changing the curve shape. This operation is enabled by the **Oslo algorithm**[40]. This algorithm inserts another set of knot values in the original knot vector, so that a curve will have a new knot vector without changing its shape. Now we can define an original curve as follows:

$$\mathbf{R}(t) = \sum_{i=0}^{n} N_{i,k}(t)\mathbf{P}_i \tag{4.16}$$

where the knot vector of this curve is

$$[x_0 \ x_1 \ \dots \ x_{n+k}]$$

and a new vector is obtained by inserting another set of knot vector values to give

$$[y_0 \ y_1 \ \dots \ y_{m+k}] \qquad (m > n)$$

Then, the B-spline curve with the same shape as the original curve becomes

$$\mathbf{S}(s) = \sum_{j=0}^{m} M_{j,k}(s)\mathbf{Q}_j \tag{4.17}$$

where $M_{j,k}(s)$ is the B-spline basis that can be defined by the new knot vector. The new control point can be obtained from

$$\mathbf{Q}_j = \sum_{i=0}^{n} \alpha_{i,j}^k \mathbf{P}_i \qquad (0 \le i \le n, 0 \le j \le m)$$

where $\alpha_{i,j}^k$ is given by

$$\alpha_{i,j}^1 = \begin{cases} 1 & \text{if } x_i \le y_j < x_{i+1} \\ 0 & \text{otherwise} \end{cases} \tag{4.18}$$

$$\alpha_{i,j}^k = \frac{y_{j+k-1} - x_i}{x_{i+k-1} - x_i}\alpha_{i,j}^{k-1} + \frac{x_{i+k} - y_{j+k-1}}{x_{i+k} - x_{i+1}}\alpha_{i+1,j}^{k-1}$$

Thus a curve can be divided by inserting multiple knot vector values in the original knot vector, where these values correspond to the order at the parameter position to be divided.

4.6 NURBS

The types of curves that can be represented in Bézier form are limited because the Bézier curve is a polynomial curve. Likewise, the B-spline curve is a polynomial curve, so that conic sections cannot be represented. To overcome this limitation, the **rational B-spline curve** was proposed. Versprille[152] was the first to discuss the rational B-spline curve and Tiller[145] and Piegl[110] developed the basis for current discussion.

4.6.1 How to represent NURBS

A rational B-spline curve is represented as

$$\mathbf{R}(t) = \frac{\sum_{i=0}^{n} N_{i,k}(t) w_i \mathbf{P}_i}{\sum_{i=0}^{n} N_{i,k}(t) w_i} \tag{4.19}$$

where $N_{i,k}(t)$ is the B-spline basis, expressed in Equation (4.13). NURBS stands for **NonUniform Rational B-Spline**. The term nonuniform describes one of the three forms of knot vectors found in B-spline curves. It means that there is no specific form for the knot vector values and a knot vector can contain any knot values once it satisfies the condition of monotonic increase. The characteristics of B-spline curves are passed on to rational B-spline curves, so that NURBS also has the B-spline characteristics of nonuniform vectors. Since a rational B-spline curve has a nonuniform knot vector, by assigning another set of values to the internal knot values, a curve can be represented as if it consisted of two or more connected curves. Even a sequence of curves with sharp corners can be regarded as one continuous curve. The above equation has enabled all types of curves to be expressed as one curve. Because of these representation abilities, NURBS was included in the IGES standard in 1983. Hence NURBS is now being introduced in various CAD/CAM systems.

4.6.2 How to represent conic sections with NURBS

A conic section is a natural quadric curve. We now define a NURBS curve by a knot vector

$$[0\ 0\ 0\ 1\ 1\ 1]$$

and having the same multiplicity as the order at the end, to have the same curve representation as that of a rational Bézier curve. Then, the method to create a conic section from a rational Bézier curve, explained in Subsection 4.4.3, can be applied here. In Figure 4.8, an arc with 90° central angle is to be created in an isosceles triangle with a base angle θ, of 45°. The weights of the three control points are:

$$w_0 = 1, \quad w_1 = \cos\theta = \frac{1}{\sqrt{2}}, \quad w_2 = 1$$

With this condition, a circle can be represented by one NURBS (curve) as shown in Figure 4.12. In the Figure, □ indicates a control point.

The knot vector of this NURBS is:

$$[0\ 0\ 0\ 1\ 1\ 2\ 2\ 3\ 3\ 4\ 4\ 4]$$

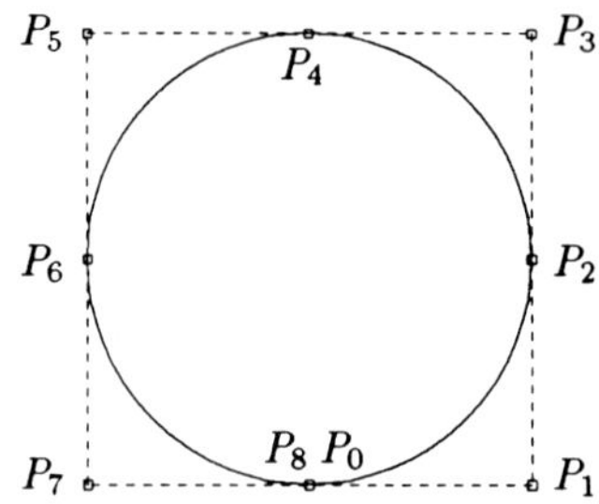

Fig. 4.12. Representation of a circle with NURBS

and the weights w_i of the control points P_i are

$$[w_0\ w_1\ w_2\ w_3\ w_4\ w_5\ w_6\ w_7\ w_8\ ] = \left[1\ \frac{1}{\sqrt{2}}\ 1\ \frac{1}{\sqrt{2}}\ 1\ \frac{1}{\sqrt{2}}\ 1\ \frac{1}{\sqrt{2}}\ 1\ \right]$$

4.6.3 Connection of curves

In this subsection, we consider the case where two curves whose control points have weights are connected. Let us take an example of two curves on a plane as shown below:

Order = 3
Control points $[x_i, y_i, w_i]$: [0,0,1], [0,1,$\frac{1}{\sqrt{2}}$], [1,1,1]
Knot vector:[0 0 0 1 1 1]

Order = 3
Control points $[x_i, y_i, w_i]$: [1,1,1], [2,1,1], [3,1.5,1], [3.5,2.5,1]
Knot vector:[0 0 0 1 2 2 2]

These two curves are connected into one curve as:

Order = 3
Control points $[x_i, y_i, w_i]$:
[0,0,1], [0,1,$\frac{1}{\sqrt{2}}$], [1,1,1], [2,1,1], [3,1.5,1], [3.5,2.5,1]
Knot vector:[0 0 0 1 1 2 3 3 3]

The end control point of one curve is connected to the start control point of the other curve. The control point positions and weights remain unchanged, hence only the knot vectors are operated.

Let us look at the operation in detail. First, add the last multiple knot value of the first curve (in this case, 1) to every knot value of the second curve like:

[0 0 0 1 1 1] [1 1 1 2 3 3 3]

Next, decrease the multiplicity of the last multiple knot value of the first curve by one and remove all of the first multiple knot values from the second curve's knot vector. Then, combine the first and second curve's knot vectors like:

$$[0\ 0\ 0\ 1\ 1\ 2\ 3\ 3\ 3]$$

With this operation, the two curves can be connected with C^0 continuity, but the curve shapes remain unchanged. In the above example, the weights of the contact control points were equal and therefore no operation on the weights was necessary. If two control points with different weights are to be connected, the weights should be leveled, i.e. first, multiply weights of every control point on one curve by a real number, and then connect the control points in the same way as above.

4.7 Programs

This section shows three programs:

(i) "curvePosition", the program to obtain the coordinates of a point on a curve by using the curve data as explained in Subsection 3.4.4 and the parameter,

(ii) "curveDeriv", the program to obtain the derivative vector of a curve from the curve data and the parameter, and

(iii) "curveDivide", the program to divide a curve into two at a specified parameter.

In these programs, the data on a point is defined by the following structure:

```
struct dbPoint {
    double x, y, z;
};
```

The data on a point with weight is defined by the following structure:

```
struct dbHPoint {
        double x, y, z, w;
};
```

The data on a vector is defined by the following structure:

```
struct dbVector {
    double x, y, z;
};
```

The data on a curve is defined by the following structure:

```
struct dbCurve {
    int     type;                       /* Type of a curve */
    union {
        struct dbLine                   ln;
        struct dbArc                    ar;
        struct dbBezier_curve           bc;
        struct dbRational_Bezier_curve  rbc;
}       *data;
};
struct dbLine {
    struct dbPoint pnt1, pnt2;      /* Start point, end point */
};
struct dbArc {
    double          rad;            /* Radius */
    struct dbPoint  cen;            /* Center */
    struct dbPoint  pnt1, pnt2;     /* Start point, end point */
};
struct dbBezier_curve {
    int             deg;            /* Degree */
    struct dbPoint  *ctlp;          /* Control point */
};
struct dbRational_Bezier_curve {
    int             deg;            /* Degree */
    struct dbHPoint *ctlp;          /* Control point */
};
```

4.7.1 Program to obtain coordinates on a curve

The program "curvePosition" is to obtain coordinates of a point on a curve using the input pointer to the curve data and the parameter.

```
/*
 * curvePosition --- Obtain coordinates of a point on a curve.
 *
 *  Parameters:
 *    cv       (I)  Curve data
 *    t        (I)  Curve parameter
 *    pnt      (O)  Point on the curve
 */
void
curvePosition(cv, t, pnt)
    struct dbCurve *cv;
    double t;
    struct dbPoint *pnt;
{
    switch (cv->type) {
    case DB_LINE :
        linePosition(&cv->data->ln, t, pnt);
```

```
        return;
    case DB_ARC :
        arcPosition(&cv->data->ar, t, pnt);
        return;
    case DB_BEZCURVE :
        bezPosition(&cv->data->bc, t, pnt);
        return;
    case DB_RBEZCURVE :
        rbezPosition(&cv->data->rbc, t, pnt);
        return;
    }
}

/*
 * linePosition --- Obtain coordinates of a point on a line.
 *
 *  Parameters:
 *    lp      (I)  Line data
 *    t       (I)  Line parameter
 *    pnt     (O)  Point on the line
 */
void
linePosition(lp, t, p)
    struct dbLine *lp;
    double t;
    struct dbPoint *p;
{
    struct dbVector vec;

    /*
     * "SubVector" is the macro to obtain the vector
     * that starts from lp->pnt1 and ends at lp->pnt2.
     * The obtained value is returned in to "vec".
     */
    SubVector(&lp->pnt2, &lp->pnt1, &vec);

    p->x = lp->pnt1.x + t * vec.x;
    p->y = lp->pnt1.y + t * vec.y;
    p->z = lp->pnt1.z + t * vec.z;
}

/*
 * arcPosition --- Obtain coordinates of a point on an arc.
 *
 *  Parameters:
 *    ap      (I)  Arc data
 *    t       (I)  Arc parameter
 *    pnt     (O)  Point on the arc
```

```
 */
void
arcPosition(ap, t, p)
    struct dbArc *ap;
    double t;
    struct dbPoint *p;
{
    double angle, rotateangle;

    /*
     * "GetArcAngle" is the function to obtain the angle
     * between the start and end points of an arc.
     * The obtained value is returned in to "angle".
     */
    GetArcAngle(ap, &angle);

    rotateangle = angle*t;

    /*
     * "RotateAngle" is the function to obtain the position
     * of a point that is rotated by "rotateangle" degrees
     * from the start point.
     * The obtained value is returned in to "p".
     */
    RotateAngle(ap, &rotateangle, p);
}

/*
 * bezPosition --- Obtain coordinates of a point on a Bezier curve.
 *
 * Parameters:
 *    bp      (I)  Bezier curve data
 *    t       (I)  Bezier curve parameter
 *    pnt     (O)  Point on the curve
 */
void
bezPosition(bp, t, pnt)
    struct dbBezier_curve *bp;
    double t;
    struct dbPoint *pnt;
{
    register double *clist, *tlist, *t1list;
    register int i;
    double tmp;

    clist = (double *) malloc((bp->deg+1) * sizeof(double));
    tlist = (double *) malloc((bp->deg+1) * sizeof(double));
    t1list = (double *) malloc((bp->deg+1) * sizeof(double));
```

```
    if (clist == 0 || tlist == 0 || t1list == 0) {
        fprintf(stderr, "malloc error\n");
        exit(1);
    }
    /*
     * Make a list of the calculation of nCi.
     */
    GetCombList(bp->deg, clist);
    /*
     * Make a list of the calculation of t^i.
     */
    GetPowerList(t, bp->deg, tlist);
    /*
     * Make a list of the calculation of (1-t)^i.
     */
    GetPowerList(1-t, bp->deg, t1list);

    p->x = p->y = p->z = 0.0;
    for (i = 0; i <= bp->deg; i++) {
        /*
         * Calculate the Bernstein basis.
         */
        tmp = clist[i] * tlist[i] * t1list[bp->deg-i];
        p->x += tmp * bp->ctlp[i].x;
        p->y += tmp * bp->ctlp[i].y;
        p->z += tmp * bp->ctlp[i].z;
    }
    free(clist);
    free(tlist);
    free(t1list);
}

/*
 * GetCombList --- Make the table list of combinations.
 *
 *  Parameters:
 *    n       (I)  Number of materials
 *    table   (O)  Table list of combinations
 */
void
GetCombList(n, table)
    int n;
    double *table;
{
    register double *p, *q, *r;
    register double *ip, *np;

    np = (ip = table) + n;
```

```
    *np = *ip = 1.0;

    while ((q = r = ++ip) < np) {
        *q-- = 1.0;
        for (p = table + 1; p < r; *p++ = (*r-- += *q))
            ;
        if  (p == r)
            *p += *p;
    }
}

/*
 * GetPowerList --- Make the list of powers of real numbers.
 *
 *  Parameters:
 *    t       (I)  Parameter
 *    n       (I)  Number
 *    table   (O)  List
 */
void
GetPowerList(t, n, table)
    double t;
    int n;
    double *table;
{
    list[0] = 1.0;
    for (i = 1; i <= n; i++) {
       list[i] = t * list[i-1];
    }
}
```

A point on a rational Bézier curve can also be obtained by the same procedure as bezPosition.

4.7.2 Program to obtain a derivative vector on a curve

The program "curveDeriv" is to obtain a derivative vector of a curve by inputting the pointer to the curve data and the parameter.

```
/*
 * curveDeriv --- Obtain derivative vector of a curve.
 *
 *  Parameters:
 *    cv      (I)  Curve data
 *    t       (I)  Curve parameter
 *    drv     (O)  Derivative vector of the curve
 */
void
```

```
curveDeriv(cv, t, drv)
    struct dbCurve *cv;
    double t;
    struct dbVector *drv;
{
    switch (cv->type) {
    case DB_LINE :
        lineDeriv(&cv->data->ln, t, drv);
        return;
    case DB_ARC :
        arcDeriv(&cv->data->ar, t, drv);
        return;
    case DB_BEZCURVE :
        bezDeriv(&cv->data->bc, t, drv);
        return;
    case DB_RBEZCURVE :
        rbezDeriv(&cv->data->rbc, t, drv);
        return;
    }
}

/*
 * lineDeriv --- Obtain a derivative vector of a line.
 *
 *  Parameters:
 *    lp      (I)  Line data
 *    t       (I)  Line parameter
 *    drv     (O)  Derivative vector of the line
 */
void
lineDeriv(lp, t, drv)
    struct dbLine *lp;
    double t;
    struct dbVector *drv;
{
    /*
     * Obtain the vector that starts from lp->pnt1
     * and ends at lp->pnt2.
     */
    SubVector(&lp->pnt2, &lp->pnt1, drv);
}

/*
 * arcDeriv --- Obtain a derivative vector of an arc.
 *
 *  Parameters:
 *    ap      (I)  Arc data
 *    t       (I)  Arc parameter
```

```
 *    drv      (O)  Derivative vector of the arc
 */
void
arcDeriv(ap, t, drv)
    struct dbArc *ap;
    double t;
    struct dbVector *drv;
{
    struct dbPoint pnt0;
    struct dbVector vec0, vec1, vec2;
    double theta;

    arcPosition(ap, t, &pnt0);
    SubVector(&pnt0, &ap->cen, &vec0);
    SubVector(&ap->pnt1, &ap->cen, &vec1);
    SubVector(&ap->pnt2, &ap->cen, &vec2);
    theta = acos(InnerProduct(&vec1, &vec2) / (ap->rad*ap->rad));
    CrossProduct(&ap->nrm, &vec0, drv);
    drv *= theta;
}

/*
 * bezDeriv --- Obtain a derivative vector of a Bezier curve.
 *
 *  Parameters:
 *    bp       (I)  Bezier curve data
 *    t        (I)  Bezier curve parameter
 *    drv      (O)  Derivative vector of the curve
 */
void
bezDeriv(bp, t, drv)
    struct dbBezier_curve *bp;
    double t;
    struct dbVector *drv;
{
    register double *clist, *tlist, *t1list;
    register int i;
    double tmp;

    clist = (double *) malloc((bp->deg) * sizeof(double));
    tlist = (double *) malloc((bp->deg) * sizeof(double));
    t1list = (double *) malloc((bp->deg) * sizeof(double));
    if (clist == 0 || tlist == 0 || t1list == 0) {
        fprintf(stderr, "malloc error\n");
        exit(1);
    }

    GetCombList(bp->deg-1, clist);
```

```
    GetPowerList(t, bp->deg-1, tlist);
    GetPowerList(1-t, bp->deg-1, t1list);

    drv->x = drv->y = drv->z = 0.0;
    for (i = 0; i < bp->deg; i++) {
        tmp = clist[i] * tlist[i] * t1list[bp->deg-i-1];
        drv->x += tmp * (bp->ctlp[i+1].x - bp->ctlp[i].x);
        drv->y += tmp * (bp->ctlp[i+1].y - bp->ctlp[i].y);
        drv->z += tmp * (bp->ctlp[i+1].z - bp->ctlp[i].z);
    }
    drv->x *= (double) bp->deg;
    drv->y *= (double) bp->deg;
    drv->z *= (double) bp->deg;
    free(clist);
    free(tlist);
    free(t1list);
}
```

A derivative vector of a rational Bézier curve can also be obtained by the same procedure as bezDeriv.

4.7.3 Program to divide a curve

The program "curveDivide" is to divide a curve into two by inputting the pointer to the curve data and the parameter.

```
/*
 * cvDivide --- Divide a curve into two.
 *
 *  Parameters:
 *    cv      (I)  Curve 1
 *    param   (I)  Point parameter
 *    cv1     (O)  Curve 1
 *    cv2     (O)  Curve 2
 */
void
curveDivide(cv, param, cv1, cv2)
    struct dbCurve *cv, *cv1, *cv2;
    real param;
{
    /*
     * Allocate the area for two curves of the same type as cv.
     */
    curveCopy(cv, cv1);
    curveCopy(cv, cv2);
    if (cv1->data == 0 || cv2->data == 0) {
        fprintf(stderr, "malloc error\n");
        exit(1);
```

```
    }

    switch(cv->type) {
    case DB_LINE:
        lineDivide(&cv->data->ln, param, &cv1->data->ln,
                                          &cv2->data->ln);
        break;

    case DB_ARC:
        arcDivide(&cv->data->ar, param, &cv1->data->ar,
                                          &cv2->data->ar);
        break;

    case DB_BEZCURVE:
        bezDivide(&cv->data->bc, param, &cv1->data->bc,
                                          &cv2->data->bc);
        break;

    case DB_RBEZCURVE:
        rbezDivide(&cv->data->rbc, param, &cv1->data->rbc,
                                          &cv2->data->rbc);
        break;
    }
}

/*
 * lineDivide --- Divide a line into two.
 *
 *  Parameters:
 *    ln       (I)  Line 1
 *    param    (I)  Parameter
 *    ln1      (O)  Line 1
 *    ln2      (O)  Line 2
 */
void
lineDivide(ln, param, ln1, ln2)
    struct dbLine *ln, *ln1, *ln2;
    real param;
{
    linePosition(ln, param, &ln1->pnt2);
    ln2->pnt1 = ln1->pnt2;
}

/*
 * arcDivide --- Divide an arc into two.
 *
 *  Parameters:
 *    ap       (I)  Arc 1
```

```
 *    param   (I)  Parameter
 *    ap1     (O)  Arc 1
 *    ap2     (O)  Arc 2
 */
void
arcDivide(ap, param, ap1, ap2)
    struct dbArc *ap, *ap1, *ap2;
    real param;
{
    struct dbPoint p1;

    arcPosition(ap, param, &ap1->pnt2);
    ap2->pnt1 = ap1->pnt2;
}

/*
 * bezDivide --- Divide a Bezier curve into two.
 *
 *  Parameters:
 *    bp      (I)  Bezier curve 1
 *    param   (I)  Parameter
 *    bp1     (O)  Bezier curve 1
 *    bp2     (O)  Bezier curve 2
 */
void
bezDivide(bp, param, bp1, bp2)
    struct dbBezier_curve *bp, *bp1, *bp2;
    double param;
{
    /*
     * de Casteljau Algorithm is applied to subdivide Bezier curve.
     *
     * Now the points p0, p1, ..., pn are the control points,
     * and t is a real number specifying the dividing point.
     * And,
     *   b_{r,i}(t) = (1-t)*b_{r-1,i}(t) + t*b_{r-1,i+1}(t)
     *
     *      where   r = 1, ..., n
     *              i = 0, ..., n-r
     *      and     b_{0,i}(t) = bi
     * Then,
     *   the intermediate coefficients b_{r,i}(t) makes next
     *   array of the points.
     *
     *      b0
     *      b1 b_{1,0}
     *      b2 b_{1,1} b_{2,0}
     *      b3 b_{1,2} b_{2,1} b_{3,0}
```

```
     *      ...........................
     *      bn b_{1,n-1} b_{2,n-2} b_{3,n-3} ... b{n,0}
     *
     * In this case,
     *   the control points of the subdivided curve 1 are
     *      b0 b_{1,0} b_{2,0} ... b_{n,0}
     *   and the control points of the another subdivided curve are
     *      b_{n,0} b{n-1,1} b_{n-2,2} ... bn
     */

    register int i, r, n;
    double _param = 1.0 - param;
    register struct dbPoint *parray;

    parray = (struct dbPoint *)
        malloc((bp->deg+1)*(bp->deg+1)*sizeof(struct dbPoint));
    if (parray == 0) {
        fprintf(stderr, "malloc error\n");
        exit(1);
    }

#define PNT(r,i)        (parray+(bp->deg+1)*(r) + i)
    /*
     * Set the initial value of the control points.
     */
    for (i = 0; i <= bp->deg; i++) {
        (*PNT(0, i)) = bp->ctlp[i];
    }

    /*
     * Apply de Casteljau algorithm.
     */
    n = bp->deg;
    for (r = 1; r <= bp->deg; r++) {
        for (i = r; i <= n; i++) {
            (*PNT(r,i)).x = _param*(*PNT(r-1,i-1)).x
                                    + param*(*PNT(r-1,i)).x;
            (*PNT(r,i)).y = _param*(*PNT(r-1,i-1)).y
                                    + param*(*PNT(r-1,i)).y;
            (*PNT(r,i)).z = _param*(*PNT(r-1,i-1)).z
                                    + param*(*PNT(r-1,i)).z;
        }
    }

    /*
     * Set the subdivided control points.
     */
    for (i = 0; i <= bp->deg; i++) {
```

```
        bp1->ctlp[i] = (*PNT(i,i));
        bp2->ctlp[i] = (*PNT(bp->deg-i, bp->deg));
    }
#undef PNT
    free(parray);
}
```

The same procedure as bezDivide can be applied to divide a rational Bézier curve.

This chapter has explained several major free-form curves used in CAD systems. Besides the curves discussed in this chapter, many others are presently under examination. Some include β-splines[6] or γ-splines[15], to which new parameters are introduced for G^2 continuity. Newer methods of representing curves, if able to provide excellent representation and ease of input into a computer, will surely be incorporated into practical CAD systems.

5. FREE-FORM SURFACES

5.1 Introduction

Surfaces have been studied as an expanded form of curves for a long time. As the pioneers of studies on surface representations, Coons and Bézier can be referred. In 1964, Coons proposed the Coons surface[44] for surface representation. It is used to define surface shapes interactively in computers. Bézier proposed the Bézier surface[11,13] in 1966, also for representing surfaces and it is used in the design of automobile bodies. Both the Coons and the Bézier surfaces, used for surface representations, create rectangular patches. Since the Coons and Bézier surfaces, a lot of research has been done to expand and generalize representations by these surfaces.

This chapter explains firstly, quadric surfaces popular in the design of machine parts. Next it explains the parametric surfaces used to represent free-form surfaces, and the Coons and the Bézier surfaces as the major representations of parametric surfaces. NURBS, which were developed by rationalizing B-spline surfaces, are also explained. When considering surface representations, one should try to determine whether smooth connection between surfaces is possible. Close attention is paid to the representations of Gregory patches and rational boundary Gregory patches, surfaces supported in DESIGNBASE, and the method of how to connect these patches. The relationship between these patches and the transformation method are explained in the last section.

5.2 Quadric Surfaces

Quadric surfaces are very popular in the design of industrial products, especially, **natural quadric surfaces** such as cylinders, spheres, or cones, which are indispensable for designing industrial machine parts. Let us briefly touch upon the representations of the quadric surfaces before getting into the main subject of free-form surfaces.

There are two ways of representing quadric surfaces: geometric representation and algebraic representation. Geometric representation of quadric surfaces features the use of geometric characteristics such as the center or radius of a circle. For example, a cylinder surface can be defined by the following three geometric characteristics: the center and the radius of the base, and the vector of the cylinder axis. Likewise, a cone surface and a sphere surface can be defined by geometric characteristics.

Algebraic representation expresses a quadric surface as follows:

$$ax^2 + by^2 + cz^2 + dxy + eyz + fzx + gx + hy + iz + j = 0 \tag{5.1}$$

where $a, b, c, d, e, f, g, h, i$, and j are real numbers.

Equation (5.1) expresses an arbitrary quadric surface in 3-dimensional space. This equation can be transformed into the coordinate system that is defined by geometric characteristics[118]. By using a proper orthogonal coordinate system, Equation (5.1) is reorganized as

$$Ax^2 + By^2 + Cz^2 = D \tag{5.2}$$

where A, B, and C are real numbers. If $A = B$ and $C = 0$, this equation expresses a cylinder; if $A = B = C > 0$ and $D > 0$, it expresses a sphere with radius $\sqrt{\frac{D}{A}}$; and if $A = B$, $\frac{C}{A} < 0$, and $D = 0$, it expresses a cone (see Figure 5.1).

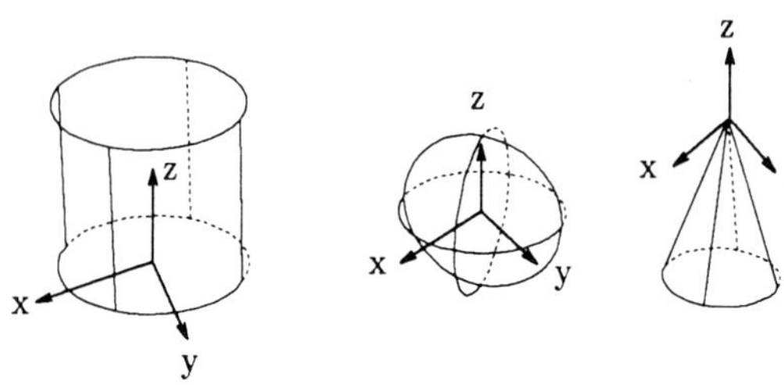

(a) Cylindrical surface (b) Spherical surface (c) Conic surface

Fig. 5.1. Natural quadric surfaces

5.3 Parametric Patches

This section explains **parametric patches**. One of the methods to input free-form surface shapes in a computer, involves the use of **feature lines**, i.e., boundary curves of surfaces and cross-section curves created by splitting a surface with a plane. Meshes created by the feature lines are called **curve meshes**. One of the methods used to model free-form surfaces, is to create curve meshes that represent surface boundaries then interpolate the surfaces using the curve meshes[32]. When representing complicated surface shapes, curve meshes often contain non-rectangular areas such as triangles or pentagons. The ability to interpolate such curve meshes with free-form surfaces is an important requirement for creating free-form surface shapes. In general, complicated surface shapes can hardly be represented with only one method of surface representation. Usually, a surface has to be divided into several surface elements and the surface elements smoothly connected, so that a free-form surface is represented. This surface element is called a **patch**.

In order to interpolate curve meshes with free-form surfaces, patches should be created from the curve meshes. The parametric patch is the standard way of representing the patch. A point on a parametric patch can be defined as a projection of the parameter $u, v\ where 0 \leq u, v \leq 1$ (see Figure 5.2).

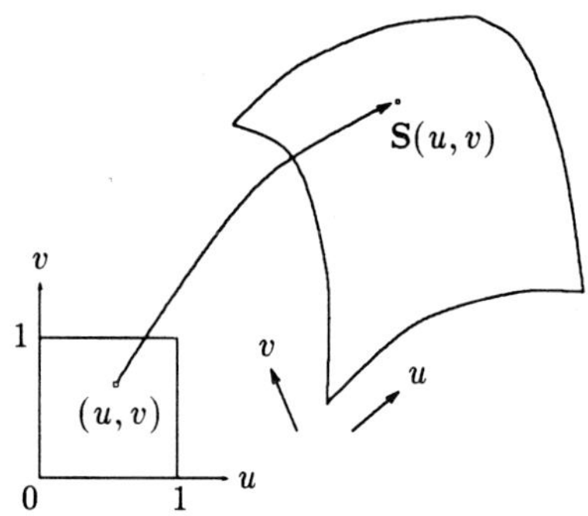

Fig. 5.2. Parametric patch

The most general, hence popular patch is a bicubic parametric patch, which means a patch expressed by a cubic expression of the parameters u and v. A bicubic parametric patch $\mathbf{S}(u, v)$ is expressed as

$$\mathbf{S}(u,v) = \begin{bmatrix} 1 & u & u^2 & u^3 \end{bmatrix} \begin{bmatrix} \mathbf{V}_{00} & \mathbf{V}_{01} & \mathbf{V}_{02} & \mathbf{V}_{03} \\ \mathbf{V}_{10} & \mathbf{V}_{11} & \mathbf{V}_{12} & \mathbf{V}_{13} \\ \mathbf{V}_{20} & \mathbf{V}_{21} & \mathbf{V}_{22} & \mathbf{V}_{23} \\ \mathbf{V}_{30} & \mathbf{V}_{31} & \mathbf{V}_{32} & \mathbf{V}_{33} \end{bmatrix} \begin{bmatrix} 1 \\ v \\ v^2 \\ v^3 \end{bmatrix} \tag{5.3}$$

where $\mathbf{V}_{ij}(i = 0, \cdots, 3; j = 0, \cdots, 3)$ is the vector that defines the surface shape. However, it is difficult to imagine the surface shape from this vector. As such, many methods have been studied to clearly relate the equation of a parametric patch with the surface shape to be created. The major methods involve the Coons surface, Bézier surface, and NURBS.

5.4 Coons Surfaces

In 1964, Coons proposed a surface representation method which defined shapes interactively with computers. This method uses rectangular areas to define a surface. A Coons surface consists of the three kinds of vectors at a corner point: a position vector, a first-order partial derivative vector, and a mixed partial derivative (see Figure 5.3).

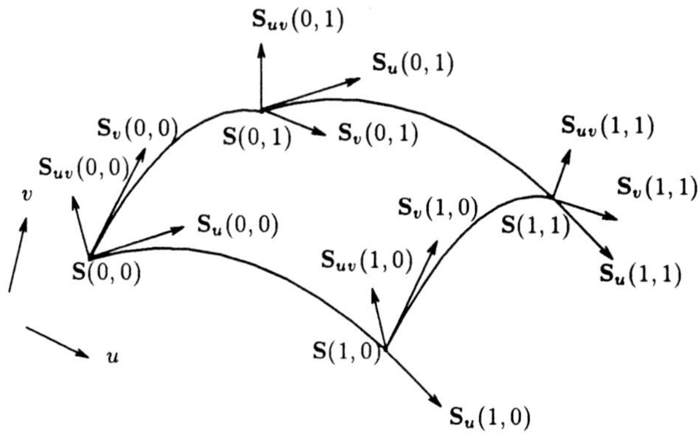

Fig. 5.3. A Coons surface

A bicubic Coons surface $\mathbf{S}(u,v) where 0 \le u, v \le 1$, is expressed as

$$\mathbf{S}(u,v) = \begin{bmatrix} h_0(u) & h_1(u) & h_2(u) & h_3(u) \end{bmatrix} \mathbf{M} \begin{bmatrix} h_0(v) \\ h_1(v) \\ h_2(v) \\ h_3(v) \end{bmatrix} \tag{5.4}$$

where

$$\begin{array}{ll} h_0(u) = (1-u)^2(2u+1), & h_1(u) = u^2(-2u+3) \\ h_2(u) = (1-u)^2 u, & h_3(u) = u^2(u-1) \end{array}$$

and the matrix $\mathbf{M}$ is

$$\mathbf{M} = \begin{bmatrix} \mathbf{S}(0,0) & \mathbf{S}(0,1) & \mathbf{S}_v(0,0) & \mathbf{S}_v(0,1) \\ \mathbf{S}(1,0) & \mathbf{S}(1,1) & \mathbf{S}_v(1,0) & \mathbf{S}_v(1,1) \\ \mathbf{S}_u(0,0) & \mathbf{S}_u(0,1) & \mathbf{S}_{uv}(0,0) & \mathbf{S}_{uv}(0,1) \\ \mathbf{S}_u(1,0) & \mathbf{S}_u(1,1) & \mathbf{S}_{uv}(1,0) & \mathbf{S}_{uv}(1,1) \end{bmatrix} \tag{5.5}$$

where $\mathbf{S}(0,0), \mathbf{S}(0,1), \mathbf{S}(1,0)$, and $\mathbf{S}(1,1)$ are the position vectors at each of the four corners of the surface. $\mathbf{S}_u(u,v)$ are the first-order partial derivatives in the u direction, $\mathbf{S}_v(u,v)$ are the first-order partial derivatives in the v direction, and $\mathbf{S}_{uv}(u,v)$ are the mixed partial derivative vectors. In summary,

$$\begin{aligned} \mathbf{S}_u(u_1,v_1) &= \left.\frac{\partial \mathbf{S}(u,v)}{\partial u}\right|_{u=u_1,v=v_1} \\ \mathbf{S}_v(u_1,v_1) &= \left.\frac{\partial \mathbf{S}(u,v)}{\partial v}\right|_{u=u_1,v=v_1} \\ \mathbf{S}_{uv}(u_1,v_1) &= \left.\frac{\partial^2 \mathbf{S}(u,v)}{\partial u \partial v}\right|_{u=u_1,v=v_1} \end{aligned}$$

The mixed partial derivative vectors at the corner points $\mathbf{S}_{uv}(u,v)$, $u = 0,1$ $v = 0,1$ are called **twist vectors**.

As is clear from Equation (5.4), the boundary curves $\mathbf{S}(u,0)$, $\mathbf{S}(u,1)$, $\mathbf{S}(0,v)$, and $\mathbf{S}(1,v)$ of a bicubic Coons surface can be expressed as cubic polynomials. For example, $\mathbf{S}(u,0)$ can be expressed as

$$\mathbf{S}(u,0) = \begin{bmatrix} h_0(u) & h_1(u) & h_2(u) & h_3(u) \end{bmatrix} \begin{bmatrix} \mathbf{S}(0,0) \\ \mathbf{S}(1,0) \\ \mathbf{S}_u(0,0) \\ \mathbf{S}_u(1,0) \end{bmatrix}$$

$\mathbf{S}_v(u,0)$, the derivative vector function that crosses a boundary curve can also be expressed as the following cubic polynomial:

$$\mathbf{S}_v(u,0) = \begin{bmatrix} h_0(u) & h_1(u) & h_2(u) & h_3(u) \end{bmatrix} \begin{bmatrix} \mathbf{S}_v(0,0) \\ \mathbf{S}_v(1,0) \\ \mathbf{S}_{uv}(0,0) \\ \mathbf{S}_{uv}(1,0) \end{bmatrix}$$

In 1967, Coons further proposed a surface representation method, which generalizes Equation (5.4)[45]. It is referred to as the **generalized Coons surface** and it is specified by boundary curves and the derivative vector functions that cross the boundary curves. The derivative vector function is called the **cross boundary derivative**[53]. Because any type of equation can be specified for the boundary curves and the cross boundary derivatives, the representation ability in the generalized Coons surface is an improvement ever the Coons surface.

A generalized Coons surface is defined by boundary curves $\mathbf{S}(u,0)$, $\mathbf{S}(u,1)$, $\mathbf{S}(0,v)$, and $\mathbf{S}(1,v)$, and the cross boundary derivatives of the boundary curves $\mathbf{S}_v(u,0)$, $\mathbf{S}_v(u,1)$, $\mathbf{S}_u(0,v)$, and $\mathbf{S}_u(1,v)$ (see Figure 5.4).

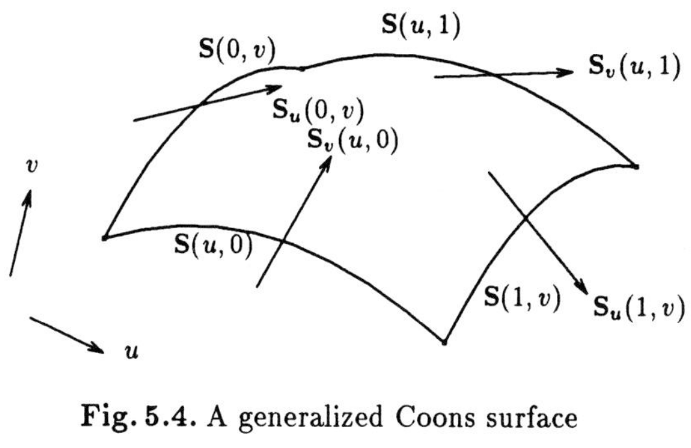

Fig. 5.4. A generalized Coons surface

A generalized Coons surface is expressed as

$$\begin{aligned}
\mathbf{S}(u,v) &= \begin{bmatrix} h_0(u) & h_1(u) & h_2(u) & h_3(u) \end{bmatrix} \begin{bmatrix} \mathbf{S}(0,v) \\ \mathbf{S}(1,v) \\ \mathbf{S}_u(0,v) \\ \mathbf{S}_u(1,v) \end{bmatrix} \\
&+ \begin{bmatrix} \mathbf{S}(u,0) & \mathbf{S}(u,1) & \mathbf{S}_v(u,0) & \mathbf{S}_v(u,1) \end{bmatrix} \begin{bmatrix} h_0(v) \\ h_1(v) \\ h_2(v) \\ h_3(v) \end{bmatrix} \\
&- \begin{bmatrix} h_0(u) & h_1(u) & h_2(u) & h_3(u) \end{bmatrix} \mathbf{M} \begin{bmatrix} h_0(v) \\ h_1(v) \\ h_2(v) \\ h_3(v) \end{bmatrix}
\end{aligned} \tag{5.6}$$

where $\mathbf{M}$ is the matrix defined in Equation (5.5).

Accordingly, Equation (5.6) can be expressed as

$$\mathbf{S}(u,v) = \mathbf{S}_A + \mathbf{S}_B - \mathbf{S}_C$$

where $\mathbf{S}_A$ indicates a cubically blended surface defined by boundary curves $\mathbf{S}(0,v)$ and $\mathbf{S}(1,v)$ and cross boundary derivatives $\mathbf{S}_u(0,v)$ and $\mathbf{S}_u(1,v)$. $\mathbf{S}_B$ indicates another surface defined by boundary curves $\mathbf{S}(u,0)$ and $\mathbf{S}(u,1)$ and cross boundary derivatives $\mathbf{S}_v(u,0)$ and $\mathbf{S}_v(u,1)$, and $\mathbf{S}_C$ indicates duplicate parts created by combining $\mathbf{S}_A$ and $\mathbf{S}_B$ (see Figure 5.5).

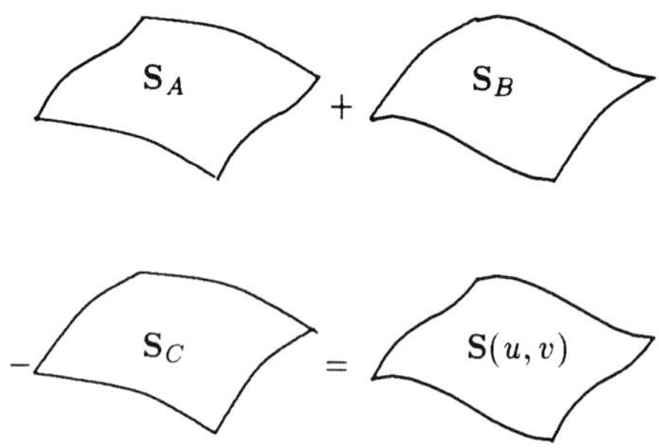

Fig. 5.5. Construction of a generalized Coons surface

When creating a patch, cross boundary derivatives may sometimes be specified in order to enable smooth connection with the adjacent patch. Theoretically, it is sufficient to specify a common cross boundary derivative to connect a patch smoothly with an adjacent patch. However, with the generalized Coons surface where cross boundary derivatives are specified at the four corner points in both the u and v parameter directions, it is necessary that the twist vector at the corner point satisfies the following conditions:

$$\left.\frac{\partial \mathbf{S}_v(u,0)}{\partial u}\right|_{u=0} = \left.\frac{\partial \mathbf{S}_u(0,v)}{\partial v}\right|_{v=0} = \mathbf{S}_{uv}(0,0)$$

$$\left.\frac{\partial \mathbf{S}_v(u,0)}{\partial u}\right|_{u=1} = \left.\frac{\partial \mathbf{S}_u(1,v)}{\partial v}\right|_{v=0} = \mathbf{S}_{uv}(1,0)$$

$$\left.\frac{\partial \mathbf{S}_v(u,1)}{\partial u}\right|_{u=0} = \left.\frac{\partial \mathbf{S}_u(0,v)}{\partial v}\right|_{v=1} = \mathbf{S}_{uv}(0,1)$$

$$\left.\frac{\partial \mathbf{S}_v(u,1)}{\partial u}\right|_{u=1} = \left.\frac{\partial \mathbf{S}_u(1,v)}{\partial v}\right|_{v=1} = \mathbf{S}_{uv}(1,1)$$

This is called the **compatibility condition**[3].

The boundary curves $\mathbf{S}(u,0), \mathbf{S}(u,1), \mathbf{S}(0,v)$, and $\mathbf{S}(1,v)$ should first be defined when creating a generalized Coons surface. After that, the twist vectors at the four corners $\mathbf{S}_{uv}(0,0)$, $\mathbf{S}_{uv}(1,0)$, $\mathbf{S}_{uv}(0,1)$, and $\mathbf{S}_{uv}(1,1)$ are defined and then, the cross boundary derivatives $\mathbf{S}_v(u,0)$, $\mathbf{S}_v(u,1)$, $\mathbf{S}_u(0,v)$, and $\mathbf{S}_u(1,v)$, in keeping with the compatibility condition. The position vectors and cross boundary derivatives can easily be conceived, but the relationship between the twist vectors and the surfaces they define cannot be intuitively understood. This is because the twist vectors $\mathbf{S}_{uv}$ are often specified to $\mathbf{0}$, otherwise a distorted surface may be created. Proper twist vectors are calculated by the methods suggested by Forrest[3] or by Adini[3]. Basically, these methods find the average twist vectors at each corner point. Thus, after making a connection a surface may be shaped differently from that imagined.

Gregory expanded the generalized Coons surface so that twist vectors could be derived from the cross boundary derivatives[62]. The elements of the mixed partial derivations $\mathbf{S}_{uv}$ in the matrix $\mathbf{M}$ is replaced as follows:

$$\mathbf{S}_{uv}(0,0) = \frac{u\dfrac{\partial \mathbf{S}_u(0,0)}{\partial v} + v\dfrac{\partial \mathbf{S}_v(0,0)}{\partial u}}{u+v}$$

$$\mathbf{S}_{uv}(0,1) = \frac{-u\dfrac{\partial \mathbf{S}_u(0,1)}{\partial v} + (v-1)\dfrac{\partial \mathbf{S}_v(0,1)}{\partial u}}{-u+v-1}$$

$$\mathbf{S}_{uv}(1,0) = \frac{(1-u)\dfrac{\partial \mathbf{S}_u(1,0)}{\partial v} + v\dfrac{\partial \mathbf{S}_v(1,0)}{\partial u}}{1-u+v}$$

$$\mathbf{S}_{uv}(1,1) = \frac{(u-1)\dfrac{\partial \mathbf{S}_u(1,1)}{\partial v} + (v-1)\dfrac{\partial \mathbf{S}_v(1,1)}{\partial u}}{u-1+v-1}$$

These four equations also satisfy the compatibility condition and thus this expansion is called **compatibility correction**.

5.5 Bézier Surfaces

A **Bézier surface**, which differs from a Coons surface, can be defined by position vectors only. These position vectors are called control points. A Bézier surface of $n \times m$ degrees is given by

$$\mathbf{S}(u,v) = \sum_{i=0}^{n}\sum_{j=0}^{m} B_i^n(u)\ B_j^m(v)\mathbf{P}_{ij} \tag{5.7}$$

where $0 \leq u$ and $v \leq 1$ and $\mathbf{P}_{ij}$ indicates the control points. $n+1$ control points lie in the u direction and $m+1$ control points lie in the v direction, totaling $(n+1) \times (m+1)$ control points for one Bézier surface. $B_i^n(u)$ and $B_j^m(v)$ are Bernstein polynomials and are given by Equation (4.3) as explained in Chapter 4.

The following explanation uses a bicubic Bézier surface as an example. As shown in Figure 5.6, a bicubic Bézier surface is represented by 16 control points $\mathbf{P}_{ij}(i = 0,\cdots,3; j = 0,\cdots,3)$.

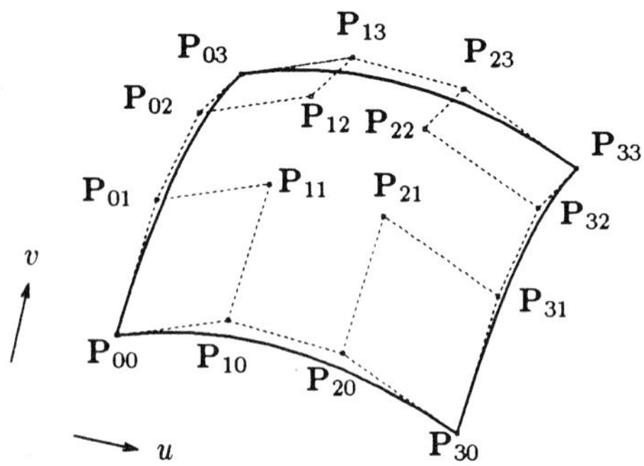

Fig. 5.6. A Bézier surface

The boundary curves $\mathbf{S}(u,0), \mathbf{S}(u,1), \mathbf{S}(0,v)$, and $\mathbf{S}(1,v)$ of a Bézier surface $\mathbf{S}(u,v)$ can be represented as Bézier curves. For example, the boundary curves $\mathbf{S}(u,0)$ of a bicubic Bézier surface $\mathbf{S}(u,v)$ are given by

$$\mathbf{S}(u,0) = \sum_{i=0}^{3} B_i^3(u)\mathbf{P}_{i0}$$

Likewise, cross boundary vectors of the boundary curves are given by

$$\mathbf{S}_v(u,0) = 3\sum_{i=0}^{3} B_i^3(u)(\mathbf{P}_{i1} - \mathbf{P}_{i0}) \tag{5.8}$$

$$\mathbf{S}_u(0,v) = 3\sum_{j=0}^{3} B_j^3(v)(\mathbf{P}_{1j} - \mathbf{P}_{0j}) \tag{5.9}$$

A Bézier surface has the following characteristics:

1. A Bézier surface is contained within the convex hull of the defining polygon net (convex hull property). Hence, the intersection of two Bézier surfaces can be roughly checked.

2. The surface follows the shape of the defining polygon net, so that the net helps the designer to easily estimate a shape.

3. A first-order derivative at an end point can be calculated from its adjacent control point. By assigning $u = 0$ in Equation (5.8) and $v = 0$ in Equation (5.9), the first-order derivatives at the corner point $\mathbf{S}(0,0)$ are obtained as

$$\begin{aligned}\mathbf{S}_u(0,0) &= 3(\mathbf{P}_{10} - \mathbf{P}_{00})\\ \mathbf{S}_v(0,0) &= 3(\mathbf{P}_{01} - \mathbf{P}_{00})\end{aligned}$$

A Bézier surface does not need twist vectors which are necessary for defining a Coons surface. The surface can intuitively be modified by just moving the control points. However, to connect Bézier surfaces smoothly, the cross boundary derivatives should be specified so as to satisfy the compatibility condition at the corner points. For example, a corner point to which four Bézier surfaces are to be connected should have the cross boundary derivatives in each of the u and v parameter directions, which satisfy the compatibility condition. To modify a surface so that there is smooth connection and the adjacent surfaces remain unchanged, attention should be paid to the cross boundary derivatives in both u and v directions when moving the control points. This largely regulates surface formation, hence a surface cannot be modified easily while satisfying the compatibility condition.

5.6 Rational Bézier Surfaces

A Bézier surface has polynomial boundary curves and therefore can be used to interpolate a surface with the curve meshes of polynomial curves. However, polynomial curves cannot represent meshes to precisely interpolate a quadratic surface. This is because the curves of a mesh have to be rational curves in order to be able to represent a quadratic surface exactly. To get around this restriction, the Bézier surface was expanded into the **rational Bézier surface**.

A rational Bézier surface of $n \times m$ degree is given by

$$\mathbf{S}(u,v) = \frac{\sum_{i=0}^{n}\sum_{j=0}^{m} B_i^n(u)B_j^m(v)w_{ij}\mathbf{P}_{ij}}{\sum_{i=0}^{n}\sum_{j=0}^{m} B_i^n(u)B_j^m(v)w_{ij}}$$

where $0 \leq u$ and $v \leq 1$, $\mathbf{P}_{ij}$ indicates the control points of a patch, w_{ij} indicates the weights of the control points, and $B_i^n(u)$ is the Bernstein polynomial given by Equation (4.3).

The following explanation uses a bicubic rational Bézier surface as an example. A bicubic rational Bézier surface is represented by 16 control points and their weights $\mathbf{P}_{ij}, w_{ij} (i = 0, \cdots, 3; j = 0, \cdots, 3)$ as shown in Figure 5.7 below.

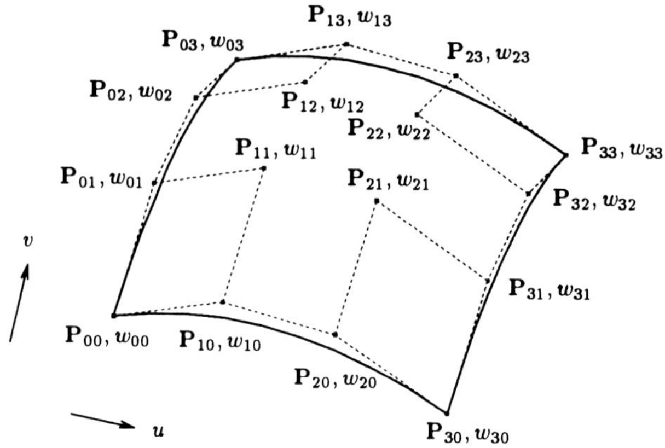

Fig. 5.7. Rational Bézier surface

The boundary curves $\mathbf{S}(u,0)$, $\mathbf{S}(u,1)$, $\mathbf{S}(0,v)$, and $\mathbf{S}(1,v)$ of a rational Bézier surface $\mathbf{S}(u,v)$ can be represented as rational Bézier curves. For example, the boundary curves $\mathbf{S}(u,0)$ of a bicubic rational Bézier surface $\mathbf{S}(u,v)$ are given by

$$\mathbf{S}(u,0) = \frac{\sum_{i=0}^{3} B_i^3(u) w_{i0} \mathbf{P}_{i0}}{\sum_{i=0}^{3} B_i^3(u) w_{i0}}$$

Likewise, the cross boundary derivatives of the boundary curves are given by

$$\begin{aligned} \mathbf{S}_v(u,0) &= \frac{3\sum_{i=0}^{3} B_i^3(u)(w_{i1}\mathbf{P}_{i1} - w_{i0}\mathbf{P}_{i0}) \sum_{i=0}^{3} B_i^3(u) w_{i0}}{\left(\sum_{i=0}^{3} B_i^3(u) w_{i0}\right)^2} \\ &\quad - \frac{3\sum_{i=0}^{3} B_i^3(u) w_{i0}\mathbf{P}_{i0} \sum_{i=0}^{3} B_i^3(u)(w_{i1} - w_{i0})}{\left(\sum_{i=0}^{3} B_i^3(u) w_{i0}\right)^2} \end{aligned} \tag{5.10}$$

The characteristics of the rational Bézier surface are as follows:

1. A rational Bézier surface, if its control points each have weight of 0 or positive, is contained within the convex hull of the defining polygon net (convex hull property). Hence, the intersection of two rational Bézier surfaces can be roughly checked.

2. As described in Section 4.4, rational Bézier curves can represent conic sections precisely. Hence rational Bézier surfaces can also precisely represent quadratic surfaces.

3. A derivative vector at an end point can be calculated from its adjacent control point and its weight. For example, by assigning $u = 0$ in Equation (5.10), the derivative vectors at the end point $\mathbf{S}(0,0)$ are obtained as

$$\mathbf{S}_u(0,0) = \frac{3w_{10}(\mathbf{P}_{10} - \mathbf{P}_{00})}{w_{00}}$$

$$\mathbf{S}_v(0,0) = \frac{3w_{01}(\mathbf{P}_{01} - \mathbf{P}_{00})}{w_{00}}$$

5.7 NURBS

As described in Section 5.5, a Bézier surface can be modified by moving its control points, but the smooth connection of two Bézier surfaces is restricted by the compatibility condition. The **B-spline surface** developed by Gordon and Riesenfeld[59] solves this problem of the connection.

It has already been shown, using the example of the Bézier surface, that a polynomial surface can be expanded into a rational surface by adding weights to its control points. By applying the same theory, a B-spline surface is generalized into the rational B-spline surface. This section focuses on NURBS, one of the rational B-spline surfaces whose knot vector is nonuniform.

Before getting into further discussion on NURBS, let us make clear how a B-spline surface is defined. A B-spline surface is given by

$$\mathbf{S}(u,v) = \sum_{i=0}^{n}\sum_{j=0}^{m} N_{i,k}(u)M_{j,l}(v)\mathbf{P}_{ij} \tag{5.11}$$

where $N_{i,k}(u)$ and $M_{j,l}(v)$ are the basis functions in the u and v parameter directions, as shown by Equation (4.13).

Equation (5.11) is rationalized, giving the **rational B-spline surface** as follows:

$$\mathbf{S}(u,v) = \frac{\sum_{i=0}^{n}\sum_{j=0}^{m} N_{i,k}(u)M_{j,l}(v)w_{ij}\mathbf{P}_{ij}}{\sum_{i=0}^{n}\sum_{j=0}^{m} N_{i,k}(u)M_{j,l}(v)w_{ij}}$$

This equation expresses a rational B-spline surface of $k \times l$ degrees (hence the order $k+1, l+1$) with the number of control points $(n+1) \times (m+1)$. w_{ij} indicates the weight of each control point. The knot vector of this surface is given by

$$[x_0\ x_1\ \cdots\ x_p]\ [y_0\ y_1\ \cdots\ y_q]$$

where the conditions $p = n + k + 1$ and $q = m + l + 1$ must be satisfied.

As described in Section 4.6, a NURBS curve can express continuous curve segments by multiplying its internal knot values. Using the same theory, a NURBS surface can express continuous patches by multiplying the knot values set in each parameter direction.

NURBS have the following characteristics:

1. A NURBS surface, if its control points each have weights of 0 or positive, is contained within the convex hull of the defining polygon net (convex hull property).

2. The maximum order in each parameter direction is smaller than the number of defining control points in that direction.

3. NURBS of order $k+1, l+1$ (if knot values are not multiples) are C^{k-1}, C^{l-1} continuous everywhere in each parameter direction.

4. Part of a NURBS surface can be modified by moving its defining control points while a whole Bézier surface is modified by the same operation.

5. The degree does not always correspond to the number of the control points; this explains the fact that the degree never changes even though new control points might be added.

6. The NURBS surface is a rational Bézier surface if the following two conditions are satisfied:

   (a) If the number of defining control points is equal to the order in each parameter direction; i.e., the number of knots is twice as many as the order.

   (b) There are no interior knot values.

   Therefore a NURBS surface can precisely represent a quadratic surface.

5.8 Gregory Patches

Gregory expanded the generalized Coons surface so that its twist vectors could be derived from the cross boundary derivatives. Based on this expansion, Chiyokura et al.[35] were able to propose a method of representing surfaces that compensated for the restrictions of the compatibility condition in the Bézier surface. This representation is called the **Gregory patch**.

A bicubic Gregory patch is represented by 20 control points $\mathbf{P}_{ijk}(i = 0,\cdots,3; j = 0,\cdots,3; k = 0,1)$ as shown in Figure 5.8. A bicubic Gregory patch is given by

$$\mathbf{S}(u,v) = \sum_{i=0}^{3}\sum_{j=0}^{3} B_i^3(u)B_j^3(v)\mathbf{Q}_{ij}(u,v) \tag{5.12}$$

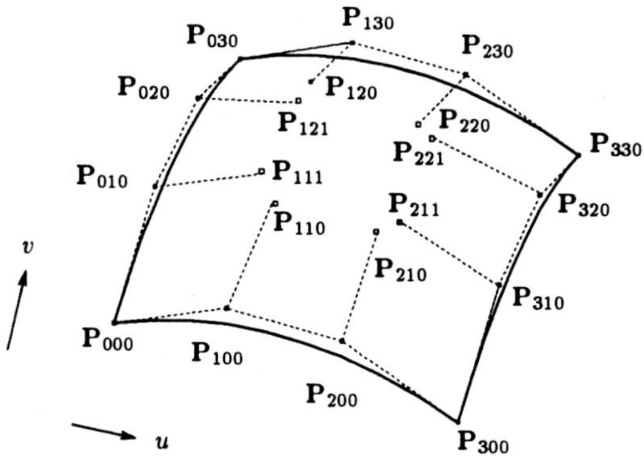

Fig. 5.8. Gregory patch

where $B_i^3(u)$ and $B_j^3(v)$ are the Bernstein bases defined by Equation (4.3). The control points $\mathbf{P}_{ijk}$ and $\mathbf{Q}_{ij}$ have the following relationships:

- If $i \neq 1,2$ or $j \neq 1,2$, then

$$\mathbf{Q}_{ij}(u,v) = \mathbf{P}_{ij0}$$

- If $i = 1,2$ and $j = 1,2$, then

$$\begin{aligned}
\mathbf{Q}_{11}(u,v) &= \frac{u\mathbf{P}_{110} + v\mathbf{P}_{111}}{u+v} \\
\mathbf{Q}_{12}(u,v) &= \frac{u\mathbf{P}_{120} + (1-v)\mathbf{P}_{121}}{u+(1-v)} \\
\mathbf{Q}_{21}(u,v) &= \frac{(1-u)\mathbf{P}_{210} + v\mathbf{P}_{211}}{(1-u)+v} \\
\mathbf{Q}_{22}(u,v) &= \frac{(1-u)\mathbf{P}_{220} + (1-v)\mathbf{P}_{221}}{(1-u)+(1-v)}
\end{aligned}$$

where $0 \leq u$ and $v \leq 1$.

This representation has the following characteristics:

1. The boundary curves of a Gregory patch are Bézier curves.
2. Like the Bézier surface, the patch has the convex hull property, i.e., the characteristic that an arbitrary point on the patch lies within the defining polygon net formed by connecting neighboring control points. Hence, the intersection of two surfaces can be roughly checked at high speed.
3. A special case of the bicubic Gregory patch is the bicubic Bézier surface. If the control points $\mathbf{P}_{110}$ and $\mathbf{P}_{111}$, $\mathbf{P}_{120}$ and $\mathbf{P}_{121}$, $\mathbf{P}_{210}$ and $\mathbf{P}_{211}$, and $\mathbf{P}_{220}$ and $\mathbf{P}_{221}$ coincide respectively, the Gregory patch is a Bézier surface.

4. The cross boundary derivatives can be defined independently for each boundary curve. For example, if $v = 0$, a cross boundary derivative is set by

$$\mathbf{S}_v(u,0) = 3\sum_{i=0}^{3} B_i^3(u)(\mathbf{P}_{i10} - \mathbf{P}_{i00}) \tag{5.13}$$

As described in Section 5.5, to connect two Bézier surfaces smoothly, the compatibility condition must be satisfied at their end points, and this creates a big restriction in shape modification. In the case of the Gregory patch however, the cross boundary derivatives can be defined independently in each of the u and v parameter direction. Therefore, in order to connect two Gregory patches smoothly, the compatibility condition must be satisfied in only one parameter direction. This characteristic greatly helps in the creation of curved surface shapes.

5.9 Rational Boundary Gregory Patches

The Gregory patch is an effective representation if the curve mesh is formed by polynomial curves. However, to represent quadratic surfaces exactly, rational curves should be used for the boundary curves. The **rational boundary Gregory patch**[36] has been proposed as the representation that allows rational curves to be used for the boundary curves.

A bicubic rational boundary Gregory patch (see Figure 5.9) is represented by 32 control points $\mathbf{P}_{ijk}(i = 0, \cdots, 3; j = 0, \cdots, 3; k = 0, 1)$. The control points $\mathbf{P}_{ij0}$ and $\mathbf{P}_{ij1}$ that lie on the boundary curves have the same coordinates but their weights differ.

A bicubic rational boundary Gregory patch is given by

$$\mathbf{G}(u,v) = \frac{\sum_{i=0}^{3}\sum_{j=0}^{3} B_i^3(u)B_j^3(v)\mathbf{Q}_{ij}(u,v)}{\sum_{i=0}^{3}\sum_{j=0}^{3} B_i^3(u)B_j^3(v)w_{ij}(u,v)} \tag{5.14}$$

where B_i^3 is the Bernstein basis defined by Equation (4.3).

Each control point and its weight are given as follows:

- If $i = 0, 1$ and $j = 0, 1$,

$$\mathbf{Q}_{ij}(u,v) = \frac{u^2\mathbf{P}_{ij0}w_{ij0} + v^2\mathbf{P}_{ij1}w_{ij1}}{u^2 + v^2}$$

$$w_{ij}(u,v) = \frac{u^2 w_{ij0} + v^2 w_{ij1}}{u^2 + v^2}$$

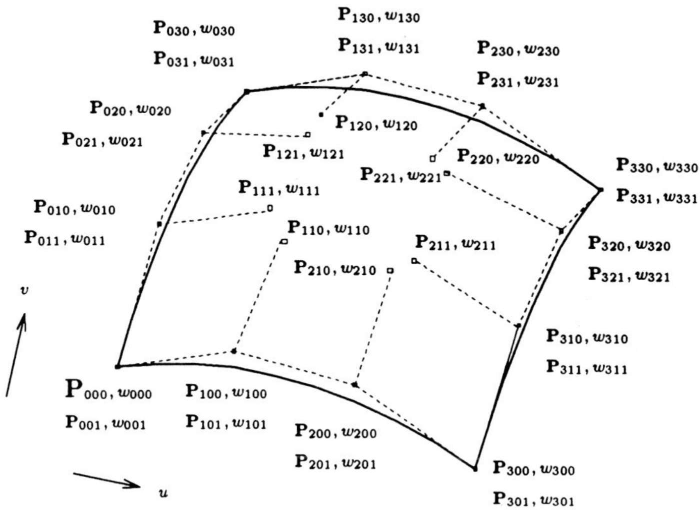

Fig. 5.9. A rational boundary Gregory patch

- If $i = 2, 3$ and $j = 0, 1$,

$$\mathbf{Q}_{ij}(u,v) = \frac{(1-u)^2\mathbf{P}_{ij0}w_{ij0} + v^2\mathbf{P}_{ij1}w_{ij1}}{(1-u)^2 + v^2}$$

$$w_{ij}(u,v) = \frac{(1-u)^2 w_{ij0} + v^2 w_{ij1}}{(1-u)^2 + v^2}$$

- If $i = 0, 1$ and $j = 2, 3$,

$$\mathbf{Q}_{ij}(u,v) = \frac{u^2\mathbf{P}_{ij0}w_{ij0} + (1-v)^2\mathbf{P}_{ij1}w_{ij1}}{u^2 + (1-v)^2}$$

$$w_{ij}(u,v) = \frac{u^2 w_{ij0} + (1-v)^2 w_{ij1}}{u^2 + (1-v)^2}$$

- If $i = 2, 3$ and $j = 2, 3$,

$$\mathbf{Q}_{ij}(u,v) = \frac{(1-u)^2\mathbf{P}_{ij0}w_{ij0} + (1-v)^2\mathbf{P}_{ij1}w_{ij1}}{(1-u)^2 + (1-v)^2}$$

$$w_{ij}(u,v) = \frac{(1-u)^2 w_{ij0} + (1-v)^2 w_{ij1}}{(1-u)^2 + (1-v)^2}$$

where $0 \leq u$ and $v \leq 1$.

This representation has the following characteristics:

1. The boundary curves of a rational boundary Gregory patch are rational Bézier curves.

2. If the weights w_{ijk} are 0 or positive, the patch has the convex hull property, as well as the properties of the Gregory patch.

3. A special case of the bicubic rational boundary Gregory patch is the bicubic rational Bézier surface. If the control points $\mathbf{P}_{110}$ and $\mathbf{P}_{111}$, $\mathbf{P}_{120}$ and $\mathbf{P}_{121}$, $\mathbf{P}_{210}$ and $\mathbf{P}_{211}$, and $\mathbf{P}_{220}$ and $\mathbf{P}_{221}$ coincide respectively, and the weights of each combination are also equal, the rational boundary Gregory patch is a rational Bézier surface. Hence, the rational boundary Gregory patch can represent a quadratic surface precisely.

4. The cross boundary derivatives can be defined independently for each boundary curve, as in the case of the Gregory patch. For example, if $w_{i00} = w_{i10}$ $(i = 0, \cdots, 3)$, the cross boundary derivatives on the boundary curves of $v = 0$ are set by

$$\mathbf{S}_v(u,0) = \frac{3\sum_{i=0}^{3} B_i^3(u) w_{i00}(\mathbf{P}_{i10} - \mathbf{P}_{i00})}{\sum_{i=0}^{3} B_i^3(u) w_{i00}} \tag{5.15}$$

   This characteristic enables curves to be created locally.

5.10 Connection of Patches

When interpolating a curve mesh with free-form surfaces, the adjacent two surfaces in the mesh are to be connected smoothly, depending on the shape of the curve mesh. The curve mesh represents features of the surface shape, hence, if a surface is created according to the feature lines, it will have the same shape that the designer imagines. For example, suppose a curve mesh consists of two surfaces to be interpolated by free-form surfaces. If curves connecting to each end point of the surfaces' common boundary curve are G^1 continuous (refer to Section 4.3), then the two surfaces should be connected to have G^1 continuity. Here, the surfaces with G^1 continuity are those whose tangent planes coincide at every point on the boundary curve of the two surfaces. In order to satisfy this condition, the tangent planes must coincide at both the end points of the boundary curve. If the tangent planes do not coincide, then the surfaces are not connected smoothly.

This section explains how to use Gregory patches and rational boundary Gregory patches to interpolate two rectangular areas lying on both sides of a boundary curve such that the resultant surface has G^1 continuity.

5.10.1 Conditions on connectivity

In order for two surfaces $\mathbf{S}^1$ and $\mathbf{S}^2$ with a common boundary curve to have G^1 continuity, the derivative vectors on the boundary curves should satisfy the condition defined by Equation (5.16). As shown in Figure 5.10, the derivative vectors that cross the boundary curve are set by $\mathbf{S}_u$ and $\mathbf{S}_v$. Then, if the equation below can be proved true at every point on the boundary curve, the two surfaces can be said to be G^1 continuous:

$$\mathbf{S}_u^2(0,v) = k(v)\mathbf{S}_u^1(1,v) + h(v)\mathbf{S}_v^1(1,v) \tag{5.16}$$

where k, and h are scalar functions of v. Equation (5.16) proves that the three derivative vectors $\mathbf{S}_u^1(1,v)$, $\mathbf{S}_v^1(1,v)$, and $\mathbf{S}_u^2(0,v)$ are coplanar at every point on the boundary curve.

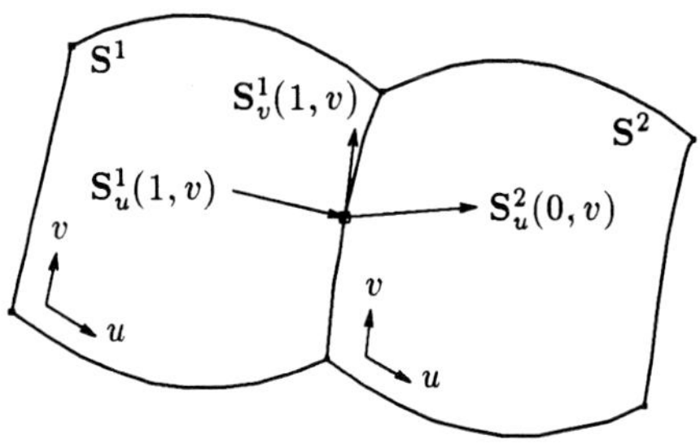

Fig. 5.10. Condition of G^1 continuity

5.10.2 Connection of Gregory patches

This subsection explains how to connect two bicubic Gregory patches so that they have G^1 continuity. Figure 5.11 shows the two Gregory patches $\mathbf{S}^1(u,v)$ and $\mathbf{S}^2(u,v)$ to be connected.

As shown in Equation (5.13), the cross boundary derivatives of a Gregory patch can be defined by the vectors between the control points on the boundary curve and their adjacent control points. If the vectors are set to $\mathbf{a}_i(i=0,\cdots,3)$, $\mathbf{b}_i(i=0,\cdots,3)$, and $\mathbf{c}_i(i=0,1,2)$, Equation (5.16) becomes

$$\sum_{i=0}^{3} B_i^3(v)\mathbf{b}_i = k(v)\sum_{i=0}^{3} B_i^3(v)\mathbf{a}_i + h(v)\sum_{i=0}^{2} B_i^2(v)\mathbf{c}_i \tag{5.17}$$

where k and h are scalar functions of v.

First, to get two patches connected with G^1 continuity, the three vectors connected to the vertices $\mathbf{V}_0$ and $\mathbf{V}_1$ should be coplanar, as proved by the following equations:

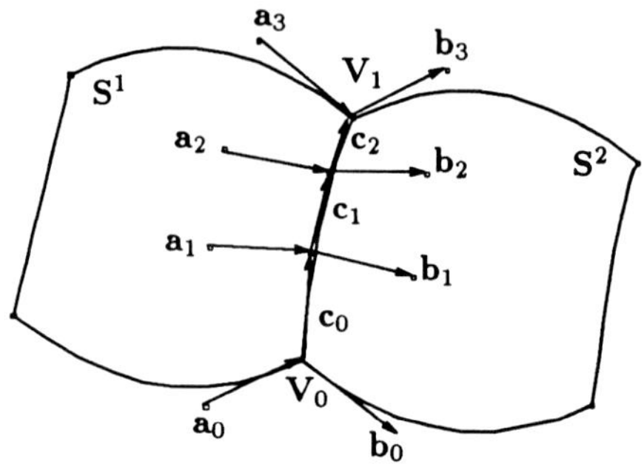

Fig. 5.11. Connection of Gregory patches

$$\mathbf{b}_0 = k_0\mathbf{a}_0 + h_0\mathbf{c}_0, \quad \mathbf{b}_3 = k_1\mathbf{a}_3 + h_1\mathbf{c}_2 \tag{5.18}$$

where k_0, k_1, h_0, and h_1 are real numbers. To satisfy Equation (5.18), the functions k and h of v are given by the following linear equations:

$$k(v) = k_0(1-v) + k_1 v$$

$$h(v) = h_0(1-v) + h_1 v$$

The function defined by the control points $\mathbf{P}_i$ is expressed by

$$\sum_{i=0}^{n} B_i^n(t)\mathbf{P}_i = \{\mathbf{P}_0, \mathbf{P}_1, \cdots, \mathbf{P}_n\}$$

Under this condition, it can be proven that

$$t\{\mathbf{P}_0, \mathbf{P}_1, \cdots, \mathbf{P}_n\} = \left\{0, \frac{\mathbf{P}_0}{n+1}, \frac{2\mathbf{P}_1}{n+1}, \cdots, \mathbf{P}_n\right\} \tag{5.19}$$

and

$$(1-t)\{\mathbf{P}_0, \mathbf{P}_1, \cdots, \mathbf{P}_n\} = \left\{\mathbf{P}_0, \frac{n\mathbf{P}_1}{n+1}, \frac{(n-1)\mathbf{P}_2}{n+1}, \cdots, \frac{\mathbf{P}_n}{n+1}, 0\right\} \tag{5.20}$$

The left-hand side of Equation (5.19) is transformed to

$$\begin{aligned} t\{\mathbf{P}_0, \mathbf{P}_1, \cdots, \mathbf{P}_n\} &= t\sum_{i=0}^{n}\binom{n}{i}t^i(1-t)^{n-i}\mathbf{P}_i \\ &= \sum_{i=0}^{n}\binom{n}{i}t^{i+1}(1-t)^{n-i}\mathbf{P}_i \end{aligned}$$

If $k = i+1$ and $m = n+1$, then

$$t\{\mathbf{P}_0, \mathbf{P}_1, \cdots, \mathbf{P}_n\} = \sum_{k=1}^{m}\binom{m-1}{k-1}t^k(1-t)^{m-k}\mathbf{P}_{k-1}$$

Since,

$$\binom{m-1}{k-1} = \frac{k}{m}\binom{m}{k}$$

therefore,

$$\begin{aligned} t\{\mathbf{P}_0, \mathbf{P}_1, \cdots, \mathbf{P}_n\} &= \sum_{k=1}^{m} \frac{k}{m}\binom{m}{k} t^k (1-t)^{m-k} \mathbf{P}_{k-1} \\ &= \left\{0, \frac{\mathbf{P}_0}{n+1}, \frac{2\mathbf{P}_1}{n+1}, \cdots, \mathbf{P}_n\right\} \end{aligned}$$

Similarly, Equation (5.20) can be proven.

Using Equations (5.19) and (5.20), the derivative vectors on the boundary curve are found to be

$$\begin{aligned} \mathbf{S}_u^1(1, v) &= 3\{\mathbf{a}_0, \mathbf{a}_1, \mathbf{a}_2, \mathbf{a}_3\} \\ \mathbf{S}_u^2(0, v) &= 3\{\mathbf{b}_0, \mathbf{b}_1, \mathbf{b}_2, \mathbf{b}_3\} \\ \mathbf{S}_v^1(1, v) &= 3\{\mathbf{c}_0, \mathbf{c}_1, \mathbf{c}_2\} \end{aligned}$$

Let us now substitute these values into Equation (5.16). Since the degree of the left-hand side is quartic, then the condition that the cross boundary derivative $\mathbf{S}_u^1$ is quadratic holds and hence the degree of the derivative is reduced by 1. So,

$$3\{\mathbf{a}_0, \mathbf{a}_1, \mathbf{a}_2, \mathbf{a}_3\} = 3\{\mathbf{a}_0, \mathbf{q}, \mathbf{a}_3\}$$

where

$$\mathbf{q} = \frac{3\mathbf{a}_1 - \mathbf{a}_0}{2} = \frac{3\mathbf{a}_2 - \mathbf{a}_3}{2}$$

Therefore, the condition of connectivity is set as

$$\{\mathbf{b}_0, \mathbf{b}_1, \mathbf{b}_2, \mathbf{b}_3\} = \{(1-v)k_0 + vk_1\}\{\mathbf{a}_0, \mathbf{q}, \mathbf{a}_3\} + \{(1-v)h_0 + vh_1\}\{\mathbf{c}_0, \mathbf{c}_1, \mathbf{c}_2\}$$

This condition is solved by using Equations (5.19) and (5.20). If the vectors $\mathbf{a}_1$ and $\mathbf{a}_2$ are such that

$$\begin{aligned} \mathbf{a}_1 &= \frac{2\mathbf{a}_0 + \mathbf{a}_3}{3} \\ \mathbf{a}_2 &= \frac{\mathbf{a}_0 + 2\mathbf{a}_3}{3} \end{aligned}$$

then $\mathbf{b}_1$ and $\mathbf{b}_2$ are calculated using the following equations:

$$\mathbf{b}_1 = \frac{(k_1 - k_0)\mathbf{a}_0}{3} + k_0\mathbf{a}_1 + \frac{2h_0\mathbf{c}_1}{3} + \frac{h_1\mathbf{c}_0}{3} \qquad (5.21)$$

$$\mathbf{b}_2 = k_1\mathbf{a}_2 - \frac{(k_1 - k_0)\mathbf{a}_3}{3} + \frac{h_0\mathbf{c}_2}{3} + \frac{2h_1\mathbf{c}_1}{3} \qquad (5.22)$$

With this procedure, the internal control points of surfaces provide G^1 continuity.

5.10.3 Connection of rational boundary Gregory patches

The procedure described in the previous subsection can be applied if connecting two rational boundary Gregory patches with G^1 continuity. As Figure 5.12 shows, the two rational boundary Gregory patches are given by $\mathbf{S}^1(u,v)$ and $\mathbf{S}^2(u,v)$ and the vectors between the control points are given by $\mathbf{a}_i$, $\mathbf{b}_i$, and $\mathbf{c}_i$. From Equation (5.15), the cross boundary derivatives can easily be calculated if the weights are properly set. The derivative vectors along the boundary curve are complicated, but the condition of connectivity as set out by Equation (5.16) is simplified by stating the end point conditions as follows:

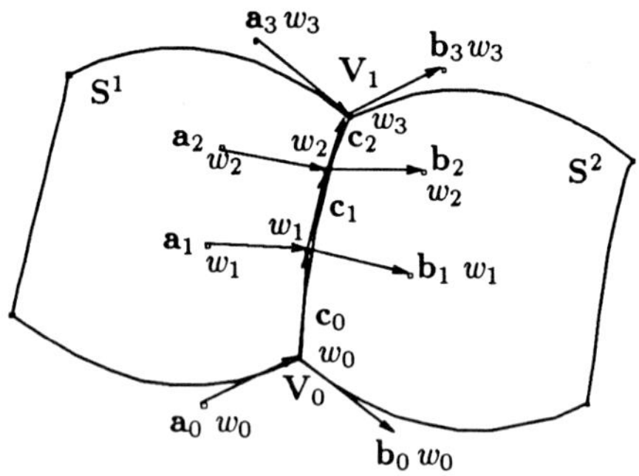

Fig. 5.12. Connection of rational boundary Gregory patches

1. If the cross boundary derivatives $\mathbf{a}_0$ and $\mathbf{b}_0$ around the vertex $\mathbf{V}_0$ are colinear as well as the cross boundary derivatives $\mathbf{a}_3$ and $\mathbf{b}_3$ around the vertex $\mathbf{V}_1$, then the equation for connection around the vertex is given by

$$\mathbf{S}^2_u(0,v) = k(v)\mathbf{S}^1_u(1,v)$$

Thus, the conditions that two patches are G^1 continuous on connection are: $\mathbf{S}^1_u$ is set by a quadratic equation and the vectors between the control points satisfy the following equations:

$$\mathbf{b}_1 = \frac{1}{w_1}\left\{k_0 w_1 \mathbf{a}_1 + \frac{(k_1 - k_0)}{3} w_0 \mathbf{a}_0\right\} \tag{5.23}$$

$$\mathbf{b}_2 = \frac{1}{w_2}\left\{k_1 w_2 \mathbf{a}_2 + \frac{(k_0 - k_1)}{3} w_3 \mathbf{a}_3\right\} \tag{5.24}$$

where w_1 and w_2 are the weights of the control points lying on the boundary curve (see Figure 5.12).

2. If the edge to which two patches $\mathbf{S}^1$ and $\mathbf{S}^2$ are connected is a non-rational curve, then the two patches can be connected in the same way as for Gregory patches.

3. If the boundary curve is a quadratic rational Bézier curve and the derivative vectors $\mathbf{a}_0$ and $\mathbf{b}_0$ around the vertex $\mathbf{V}_0$ are not colinear and neither are the cross boundary derivatives $\mathbf{a}_2$ and $\mathbf{b}_4$ around the vertex $\mathbf{V}_1$ (see Figure 5.13), then according to Equation (5.16), the equation for connection can be written as

$$\sum_{i=0}^{4} B_i^4(v) w_i \mathbf{b}_i = k(v) \sum_{i=0}^{2} B_i^2(v) w_i \mathbf{a}_i + h(v) \frac{\sum_{i=0}^{3} B_i^3(v) \mathbf{d}_i}{\sum_{i=0}^{2} B_i^2(v) w_i}$$

where

$$\begin{aligned}
\mathbf{d}_0 &= w_0 w_1 (\mathbf{P}_1 - \mathbf{P}_0) \\
\mathbf{d}_1 &= w_0 w_1 (\mathbf{P}_1 - \mathbf{P}_0) + w_0 w_2 (\mathbf{P}_2 - \mathbf{P}_0) \\
\mathbf{d}_2 &= w_0 w_2 (\mathbf{P}_2 - \mathbf{P}_0) + w_1 w_2 (\mathbf{P}_2 - \mathbf{P}_1) \\
\mathbf{d}_3 &= w_1 w_2 (\mathbf{P}_2 - \mathbf{P}_1)
\end{aligned}$$

$\mathbf{P}_0$, $\mathbf{P}_1$, and $\mathbf{P}_2$ indicate the control points on the boundary curve and w_0, w_1, and w_2 indicate their weights.

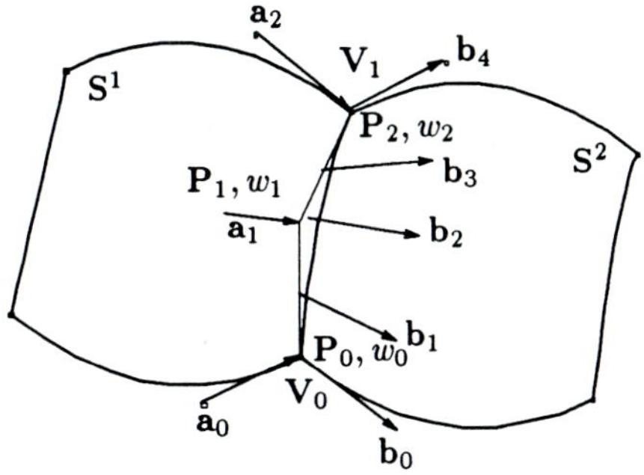

Fig. 5.13. Connection when cross boundary derivatives are not coplanar

Here, the functions $k(v)$ and $h(v)$ are set as follows:

$$\begin{aligned}
k(v) &= k_0(1-v) + k_1 v \\
h(v) &= (h_0(1-v) + h_1 v) \sum_{i=0}^{2} B_i^2(v) w_i
\end{aligned}$$

Thus, G^1 continuity can be established by obtaining the vectors between the control points that satisfy the following equations:

$$\mathbf{b}_1 = ((k_0 + k_1) w_0 \mathbf{a}_0 + 2k_0 w_1 \mathbf{a}_1 + h_1 \mathbf{d}_0 + 3h_0 \mathbf{d}_1)/4w_1 \tag{5.25}$$

$$\mathbf{b}_2 = (k_1 w_0 \mathbf{a}_0 + 4(k_0 + k_1) w_1 \mathbf{a}_1 + k_0 w_2 \mathbf{a}_2 + 3h_1 \mathbf{d}_1 + 3h_0 \mathbf{d}_2)/6w_2 \tag{5.26}$$

$$\mathbf{b}_2 = (2k_1 w_1 \mathbf{a}_1 + (k_0 + k_1) w_2 \mathbf{a}_2 + 3h_1 \mathbf{d}_2 + h_0 \mathbf{d}_3)/4w_3 \tag{5.27}$$

So far, if the boundary curve is the rational curve of degree n and the cross boundary derivatives of the boundary curve are not coplanar, then a patch of degree $2n$ is created.

These steps have proven that two rational boundary Gregory patches can be connected and be G^1 continuous.

5.11 Interpolation of Patches

The previous section showed how to interpolate two rectangular areas lying on either side of a boundary curve by using Gregory patches or rational boundary Gregory patches so that they had G^1 continuity. However, if a complicated shape is created from free-form surfaces, its curve mesh may contain areas such as triangles or pentagons which are enclosed by three or five curves. Such n-sided curve meshes should be interpolated by free-form surfaces. This section explains how to interpolate n-sided curve meshes by using Gregory patches or rational boundary Gregory patches.

5.11.1 Irregular meshes

The **regular mesh** is a mesh whose faces are rectangular, each of its nodes is connected to four faces, and the ratio of the length of the derivative vectors on each curve is constant. Otherwise, the mesh is referred to as an **irregular mesh** (see Figure 5.14).

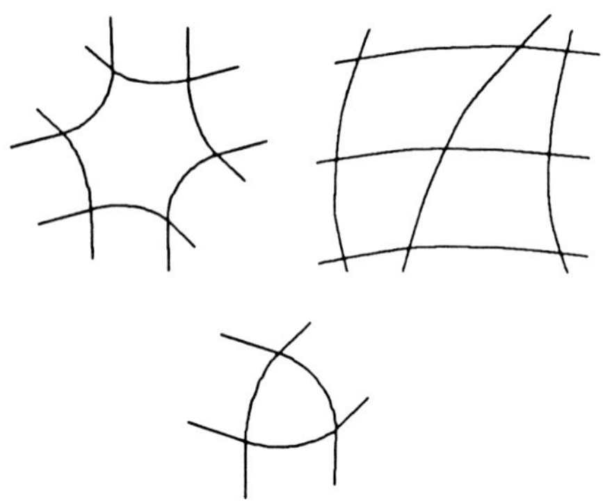

Fig. 5.14. Irregular meshes

A mesh where two faces F_3 and F_4 are connected to one face F_1 often appears when modeling shapes (see Figure 5.15(b)). Such a curve mesh is created by first splitting the boundary curve of the faces F_1 and F_2 as shown in Figure 5.15(a) and then splitting the face F_2 into the two faces F_3 and F_4. The point $\mathbf{V}_1$ to which the three faces F_1, F_3, and F_4 are connected is called a **T-node**[34].

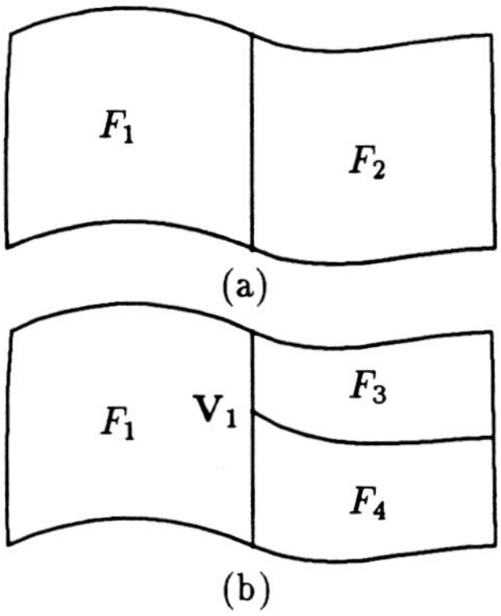

Fig. 5.15. Curve mesh with a T-node

In general, a shape can hardly be modeled as conceived, by using only regular meshes[34]. In order to overcome this restriction, irregular meshes, as well as regular meshes, should normally be interpolated by free-form surfaces.

In DESIGNBASE, if a face of a curve mesh is rectangular, even though the mesh is irregular, the face is interpolated by one Gregory patch or a rational boundary Gregory patch. If a face is non-rectangular, the face is divided into several rectangular patches and then interpolated.

5.11.2 Interpolation of curve meshes

To interpolate a curve mesh by a Gregory patch or a rational boundary Gregory patch, the cross boundary derivatives are first defined, then the control points of the patch are calculated according to the cross boundary derivatives. How the cross boundary derivatives are defined is important in the case where patches are to be interpolated so that they are G^1 continuous. The following is an explanation of how a curve mesh is interpolated:

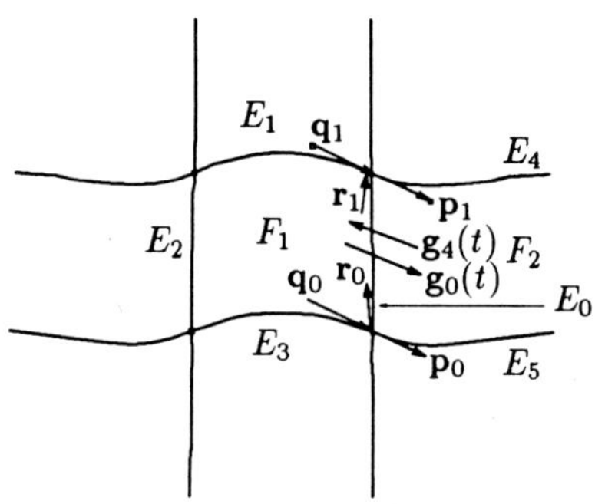

Fig. 5.16. Interpolation of a curve mesh

1. **Defining cross boundary derivatives**

   The face F_1 of Figure 5.16 is enclosed with four curves $E_i (i = 0, \cdots, 3)$.

   $\mathbf{p}_0$ is the vector between one end point of the edge $\mathbf{E}_0$ and its adjacent control point of the edge $\mathbf{E}_5$ and $\mathbf{q}_0$ is the vector between the same end point and its adjacent control point of the edge $\mathbf{E}_3$. $\mathbf{r}_0$ is also the vector between that end point and its adjacent control point of the edge $\mathbf{E}_0$. Likewise, $\mathbf{p}_1$ is the vector between the other end point of the edge $\mathbf{E}_0$ and its adjacent control point of the edge $\mathbf{E}_4$ and $\mathbf{q}_1$ is the vector between the same end point and its adjacent control point of the edge $\mathbf{E}_1$. $\mathbf{r}_1$ is also the vector between that end point and its adjacent control point of the edge $\mathbf{E}_0$.

   To interpolate a rectangular face by a Gregory patch or a rational boundary Gregory patch, the cross boundary derivatives $\mathbf{g}_i(t), (i = 0, \cdots, 3)$ that cross each boundary curve should first be defined. We shall define the cross boundary derivative of the boundary curve E_0 in Figure 5.16 to be $\mathbf{g}_0(t)$. This cross boundary derivative can be defined by the boundary curves E_0, E_1, and E_3. If the faces F_1 and F_2 are to be G^1 continuous, the cross boundary derivatives are defined by E_0, E_1, E_3, E_4, and E_5. Here, the cross boundary derivatives $\mathbf{g}_0(t)$ and $\mathbf{g}_4(t)$ defined by the faces F_1 and F_2 respectively will have the relationship expressed by $\mathbf{g}_0(t) = -\mathbf{g}_4(t)$. According to Equations (5.13) and (5.15), the cross boundary derivative is expressed as a quadratic Bézier polynomial or quadratic rational Bézier polynomial. Thus, if the faces F_1 and F_2 in Figure 5.16 are to be G^1 continuous, the cross boundary derivative is defined such that

$$\mathbf{g}_0(t) = \sum_{i=0}^{2} B_i^2(t)\mathbf{a}_i \tag{5.28}$$

   or

$$\mathbf{g}_0(t) = \sum_{i=0}^{2} B_i^2(t) w_i \mathbf{a}_i \tag{5.29}$$

   where

$$\mathbf{a}_0 = \frac{\mathbf{p}_0 + \mathbf{q}_0}{\mid \mathbf{p}_0 + \mathbf{q}_0 \mid}, \qquad \mathbf{a}_2 = \frac{\mathbf{p}_1 + \mathbf{q}_1}{\mid \mathbf{p}_1 + \mathbf{q}_1 \mid}$$

   and $\mathbf{a}_1$ is interpolated by $\mathbf{a}_0$ and $\mathbf{a}_2$.

2. **Defining internal control points**

   Because the cross boundary derivative on a boundary curve of the face F_1 is at most quadratic, it can be defined by the Bézier polynomial of the three vectors $\mathbf{a}_0$, $\mathbf{a}_1$, and $\mathbf{a}_2$. By using these three vectors and the equations for connection (see previous section), the vectors between the control points that define the internal control points can be obtained. If the face F_1 can be represented by one patch here, then the control points obtained directly represent the patch and thus the processing should be terminated.

3. **Dividing into rectangular faces**
   If a face is non-rectangular, as shown in Figure 5.14, it should be divided into several rectangular faces. After that, each rectangular face should be interpolated by a Gregory patch or a rational boundary Gregory patch.

5.11.3 Interpolation of triangular surfaces

Figure 5.17(a) illustrates internal control points of a triangular surface obtained from the equation for connection about the boundary curves. By using these control points, three internal curves are created which divide the triangular surface into three rectangular patches. The procedure is as follows:

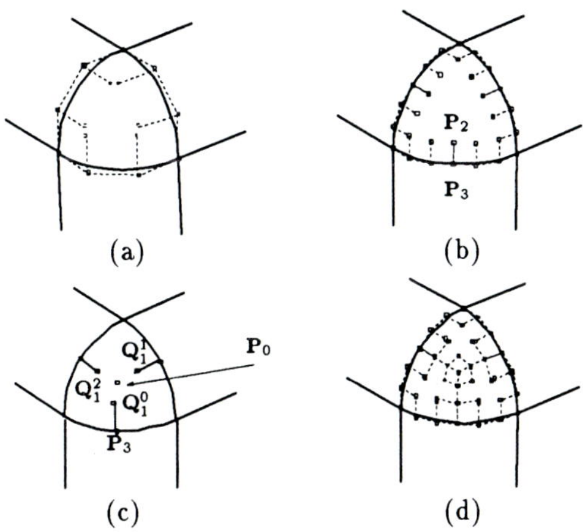

Fig. 5.17. Interpolation of a triangular surface

1. Split each of three boundary curves into two at the 0.5 parameter and also split the derivative vectors that cross boundary curves (see Figure 5.17(b)).

2. Create internal curves. According to Figure 5.17(b), the end point $\mathbf{P}_3$ of an internal curve and the control point $\mathbf{P}_2$ that defines the derivative direction at the end point are obtained. If we suppose an internal curve to be a quadratic Bézier curve, then, the control point $\mathbf{Q}_1$ is given by

$$\mathbf{Q}_1 = \frac{3\mathbf{P}_2 - \mathbf{P}_3}{2}$$

   Here, if we define the average of each $\mathbf{Q}_1^i (i = 0, 1, 2)$ to be the end point $\mathbf{P}_0$ of the internal curves, a quadratic Bézier curve is created from the three points $\mathbf{P}_0$, $\mathbf{Q}_1$, and $\mathbf{P}_3$ (Figure 5.17(c)). By elevating the degree of this Bézier curve by 1, a cubic Bézier curve is created as an internal curve.

3. Create internal control points from a curve mesh (see Figure 5.17(d)).

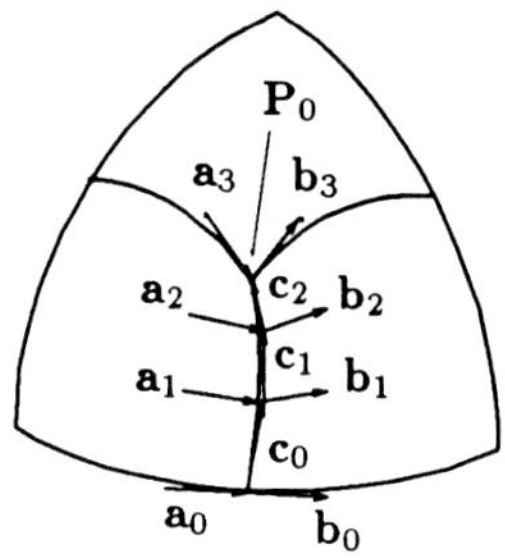

Fig. 5.18. Vectors between control points

As Figure 5.18 shows, the three vectors $\mathbf{a}_0$, $\mathbf{b}_0$, and $\mathbf{c}_0$ at an end point are coplanar and $\mathbf{a}_0$ and $\mathbf{b}_0$ make a straight line. Therefore, from Equation (5.18) h_0 becomes 0, and thus,

$$\mathbf{b}_0 = k_0 \mathbf{a}_0$$

is obtained. Likewise, $\mathbf{a}_3$, $\mathbf{b}_3$, and $\mathbf{c}_2$ are also coplanar, so that k_1 and h_1 exist to satisfy

$$\mathbf{b}_3 = k_1 \mathbf{a}_3 + h_1 \mathbf{c}_2$$

Here, $k_0 = 1$, since $\mathbf{a}_0$ and $\mathbf{b}_0$ were originally one curve which had been split at the 0.5 parameter. $k_1 = 1$ and $h_1 = 1$ are also found because the end point $\mathbf{P}_0$ is set by the average of $\mathbf{Q}_1^i$. Therefore, Equations (5.21) and (5.22) yield the following relational equations:

$$\mathbf{b}_1 = \mathbf{a}_1 + \frac{\mathbf{c}_0}{3} \tag{5.30}$$

$$\mathbf{b}_2 = \mathbf{a}_2 + \frac{2\mathbf{c}_1}{3} \tag{5.31}$$

5.11.4 Interpolation of surfaces containing an odd number of sides

When interpolating a surface containing an odd number of sides, the procedure for interpolating triangular surfaces can be applied up to the creation of internal curves. Let us take a pentagon for example. The procedure would be as follows:

1. Create internal curves in the same way as described in the previous subsection.

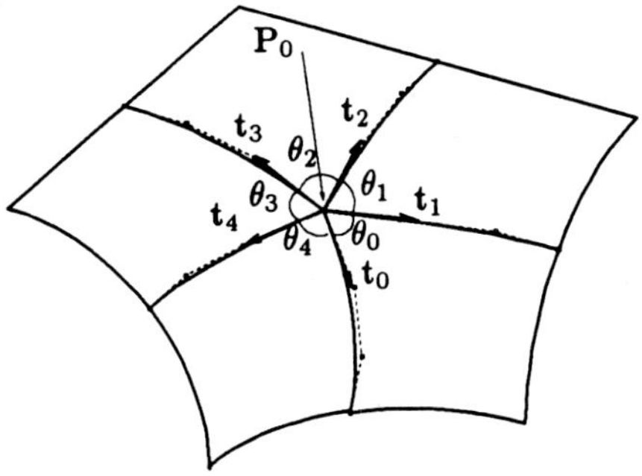

Fig. 5.19. Internal curves of a pentagon

2. Modify the internal curves created. As shown in Figure 5.19, we now define the angles of the vectors between the control points and the end point $\mathbf{P}_0$ of the internal curves, to be $\theta_i, (i = 0, \cdots, 4)$. On patches the amount of twisting at the end point $\mathbf{P}_0$ on both sides of an internal curve should coincide. Hence, setting the vectors connected to $\mathbf{P}_0$ to $\mathbf{t}_i$ and setting the angle between $\mathbf{t}_i$ and $\mathbf{t}_{i+1}$ to be θ_i,

$$| \mathbf{t}_{i+2} | \sin \theta_{i+1} = | \mathbf{t}_i | \sin \theta_i \tag{5.32}$$

should be satisfied. Then the control points of the internal curves adjacent to $\mathbf{P}_0$ are moved so as to satisfy Equation (5.32).

3. Create internal control points from the curve mesh. As Figure 5.20 shows, the vectors $\mathbf{a}_0$, $\mathbf{b}_0$, and $\mathbf{c}_0$ at the end point are coplanar. Moreover $\mathbf{a}_0$ and $\mathbf{b}_0$ make a straight line, hence the equation

$$\mathbf{b}_0 = k_0 \mathbf{a}_0 \tag{5.33}$$

is obtained. Likewise, the vectors $\mathbf{a}_3$, $\mathbf{b}_3$, and $\mathbf{c}_2$ are coplanar, thus k_1 and h_1 exist to satisfy the equation

$$\mathbf{b}_3 = k_1 \mathbf{a}_3 + h_1 \mathbf{c}_2 \tag{5.34}$$

Here, $k_0 = 1$, since $\mathbf{a}_0$ and $\mathbf{b}_0$ were originally one curve which was split into two at the 0.5 parameter. Supposing $k_1 = 1$, then Equations (5.21) and (5.22) yield the following relationships:

$$\mathbf{b}_1 = \mathbf{a}_1 + \frac{h_1 \mathbf{c}_0}{3} \tag{5.35}$$

$$\mathbf{b}_2 = \mathbf{a}_2 + \frac{2h_1 \mathbf{c}_1}{3} \tag{5.36}$$

Let us suppose the vector between the control points of adjacent boundary curves is $\mathbf{a}_2^{i+1}$ (see Figure 5.20). If the control points obtained by vectors $\mathbf{a}_2^{i+1}$ and $\mathbf{b}_2$ have common coordinates, then

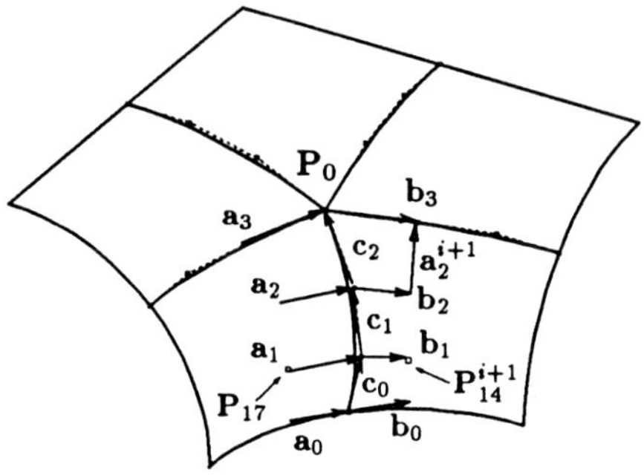

Fig. 5.20. Creation of internal control points

$$\mathbf{a}_2^{i+1} + \mathbf{b}_2 = \mathbf{c}_2 + \mathbf{b}_3 \tag{5.37}$$

is obtained. Set the control points defined by $\mathbf{a}_1$ and $\mathbf{b}_1$ to $\mathbf{P}_{17}$ and $\mathbf{P}_{14}^{i+1}$ respectively, such that

$$\mathbf{a}_1 + \mathbf{b}_1 = \mathbf{P}_{17} + \mathbf{P}_{14}^{i+1} \tag{5.38}$$

Equations (5.33) through (5.38) yield a simultaneous equation. By solving this, the internal control points can be defined (see Figure 5.21).

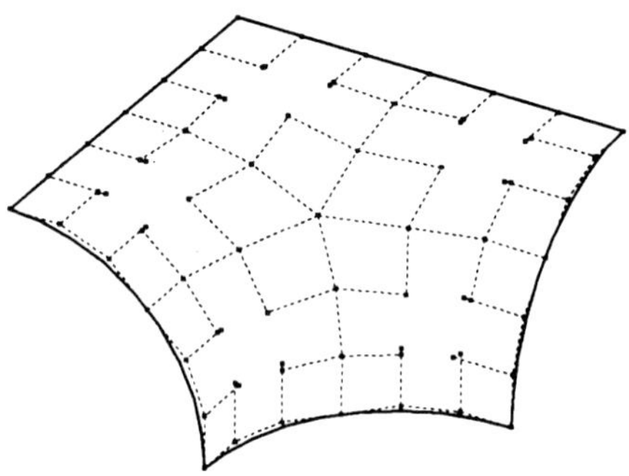

Fig. 5.21. Interpolation of a pentagon

5.11.5 Interpolation of surfaces containing an even number of sides

When interpolating a surface containing an even number of sides, the procedure for interpolating triangular surfaces can be applied up to the creation of internal curves. Since the number of internal curves is even, Equation (5.32), used in the previous subsection, cannot be applied here. In this case, therefore, each rectangular patch is interpolated after the internal curves are created.

5.11.6 Interpolation of curve meshes containing T-nodes

As illustrated in Figure 5.22, the three faces F_1, F_3, and F_4 are connected to the T-node $\mathbf{V}$. The cross boundary derivatives of the boundary curves E_1 and E_2 are controlled by the derivative vectors of the boundary curve E_3. Thus, the cross boundary derivatives cannot be defined by the method explained in Subsection 5.11.2.

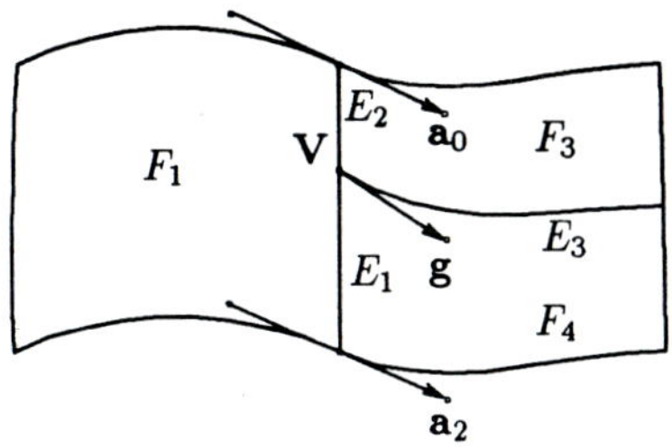

Fig. 5.22. Curve mesh containing a T-node

Consider the curve mesh in Figure 5.22, if the boundary curves E_1 and E_2 are regarded as one curve E, then the cross boundary derivative at the parameter t_0 at the point $\mathbf{V}$ on E is the derivative vector $\mathbf{g}$ of the boundary curve E_3. Hence, the cross boundary derivative $\mathbf{g}(t)$ can be defined. That is, of course, if the cross boundary derivatives at an end point $\mathbf{a}_0$ and $\mathbf{a}_2$ are already obtained. Then, according to Equation (5.28),

$$\mathbf{a}_1 = \frac{\mathbf{g} - (B_0(t_0)\mathbf{a}_0 + B_2(t_0)\mathbf{a}_2)}{B_1(t_0)}$$

can give the cross boundary vector at the T-node. If the boundary curves are non-rational, the Equations (5.21) and (5.22) for interpolating surfaces can be used to obtain the control points. If the boundary curves are rational, then Equations (5.23) and (5.24), or Equations (5.25), (5.26), and (5.27) can be used (see Figure 5.23).

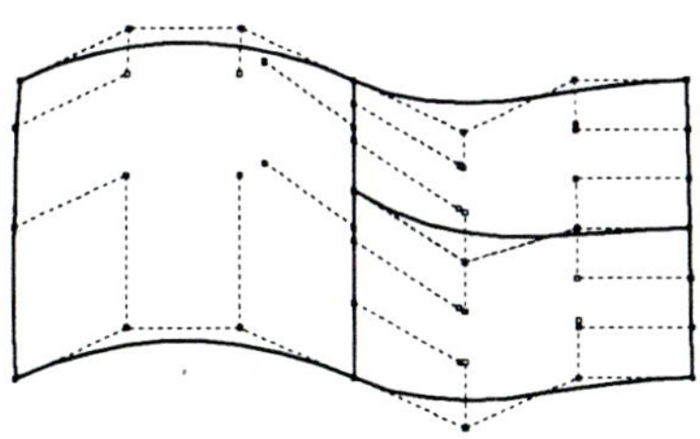

Fig. 5.23. Interpolation a curve mesh containing a T-node

5.12 Transformation of Patches

CAD systems today use various representations of surfaces depending on the purpose for which the system is being used. Some CAD systems use the Bézier surface to represent surfaces and some use NURBS. To exchange model data between CAD systems that use different representational forms, one representation must be transformed into the other.

Section 5.8 stated that a special case of the Gregory patch is the Bézier surface, which can be transformed into the Coons surface or the B-spline surface. The Gregory patch and the rational boundary Gregory patch are useful for connecting surfaces smoothly, but data exchange can sometimes be difficult, because these patches have higher representational ability than other forms of surface representations, This section explains the following:

(i) how to transform a Bézier surface into a Coons surface,

(ii) how to transform a bicubic Gregory patch into a bi-7th degree rational Bézier surface, and

(iii) how to transform a bicubic rational boundary Gregory patch into a bi-11th degree rational Bézier surface[142].

As explained in Section 5.7, a rational Bézier surface can easily be changed into a NURBS. If a Gregory patch or a rational boundary Gregory patch is first transformed into a rational Bézier surface then into a NURBS, the data on these surfaces can be sent to other CAD systems through IGES.

5.12.1 Transformation between Bézier surface and Coons surface

A bicubic Bézier surface $\mathbf{S}(u, v)$ is expressed by Equation (5.3) as follows:

$$\mathbf{S}(u,v) = \begin{bmatrix} 1 & u & u^2 & u^3 \end{bmatrix} \mathbf{M}_B \mathbf{P} \mathbf{M}_B^T \begin{bmatrix} 1 \\ v \\ v^2 \\ v^3 \end{bmatrix} \tag{5.39}$$

where

$$\mathbf{M}_B = \begin{bmatrix} 1 & 0 & 0 & 0 \\ -3 & 3 & 0 & 0 \\ 3 & -6 & 3 & 0 \\ -1 & 3 & -3 & 1 \end{bmatrix}, \mathbf{P} = \begin{bmatrix} \mathbf{P}_{00} & \mathbf{P}_{01} & \mathbf{P}_{02} & \mathbf{P}_{03} \\ \mathbf{P}_{10} & \mathbf{P}_{11} & \mathbf{P}_{12} & \mathbf{P}_{13} \\ \mathbf{P}_{20} & \mathbf{P}_{21} & \mathbf{P}_{22} & \mathbf{P}_{23} \\ \mathbf{P}_{30} & \mathbf{P}_{31} & \mathbf{P}_{32} & \mathbf{P}_{33} \end{bmatrix}$$

Likewise, a bicubic Coons surface is expressed as

$$\mathbf{S}(u,v) = \begin{bmatrix} 1 & u & u^2 & u^3 \end{bmatrix} \mathbf{M}_C \mathbf{S} \mathbf{M}_C^T \begin{bmatrix} 1 \\ v \\ v^2 \\ v^3 \end{bmatrix} \tag{5.40}$$

where

$$\mathbf{M}_C = \begin{bmatrix} 1 & 0 & 0 & 0 \\ 0 & 0 & 1 & 0 \\ -3 & 3 & -2 & -1 \\ 2 & -2 & 1 & 1 \end{bmatrix}, \mathbf{S} = \begin{bmatrix} \mathbf{S}(0,0) & \mathbf{S}(0,1) & \mathbf{S}_v(0,0) & \mathbf{S}_v(0,1) \\ \mathbf{S}(1,0) & \mathbf{S}(1,1) & \mathbf{S}_v(1,0) & \mathbf{S}_v(1,1) \\ \mathbf{S}_u(0,0) & \mathbf{S}_u(0,1) & \mathbf{S}_{uv}(0,0) & \mathbf{S}_{uv}(0,1) \\ \mathbf{S}_u(1,0) & \mathbf{S}_u(1,1) & \mathbf{S}_{uv}(1,0) & \mathbf{S}_{uv}(1,1) \end{bmatrix}$$

From Equations (5.39) and (5.40), we get

$$\mathbf{M}_B\mathbf{P}\mathbf{M}_B^T = \mathbf{M}_C\mathbf{S}\mathbf{M}_C^T$$

Hence the Bézier surface and the Coons surface can be transformed from one to the other.

5.12.2 Transformation of rational boundary Gregory patch into rational Bézier surface

First, let us recall the expression for the Gregory patch and the procedure for transforming it to that of the rational boundary Gregory patch. The denominators of $\mathbf{Q}_{ij}(u,v)$ in Equation (5.12) are multiplied by $W(u,v)$, where,

$$W(u,v) = (u+v)((1-u)+v)(u+(1-v))((1-u)+(1-v)) \tag{5.41}$$

If both sides of Equation (5.12) are multiplied by $W(u,v)$, the terms of the denominator on the right-hand side of Equation (5.12) are cleared. Here,

$$\mathbf{R}(u,v) \stackrel{\text{def}}{=} W(u,v)\mathbf{S}(u,v) \tag{5.42}$$

is set, and since this equation is the 7th degree polynomial about u and v, it can be rewritten as

$$\mathbf{R}(u,v) = \sum_{i=0}^{7}\sum_{j=0}^{7} F_i(u)F_j(v)\mathbf{P}'_{ij} \tag{5.43}$$

where $F_i(u)$ is the ith degree polynomial of u, $F_j(v)$ is the jth degree polynomial of v, and $\mathbf{P}'_{ij}$ is a vector that can be defined by the control points of a Gregory patch and the expressions of $F_i(u)$ and $F_j(v)$.

From to Equations (5.42) and (5.43),

$$\mathbf{S}(u,v) = \frac{\sum_{i=0}^{7}\sum_{j=0}^{7} F_i(u)F_j(v)\mathbf{P}'_{ij}}{W(u,v)} \tag{5.44}$$

is obtained. The numerator of Equation (5.44) is the 7th degree expression about u and v and the denominator is the quartic expression. Using the equation

$$1 = u + (1-u) = v + (1-v)$$

to adjust the denominator, it becomes a 7th degree homogeneous expression about u, $(1-u)$, v, and $(1-v)$. Then by multiplying the coefficients of the control points, in such a way that the denominator has the 7th degree Bernstein bases $B_i^7(u)$ and $B_j^7(v)$, the Equation (5.44) can be rewritten as

$$\mathbf{S}(u,v) = \frac{\sum_{i=0}^{7}\sum_{j=0}^{7} B_i^7(u)B_j^7(v)w_{ij}\mathbf{Q}_{ij}}{\sum_{i=0}^{7}\sum_{j=0}^{7} B_i^7(u)B_j^7(v)w_{ij}} \tag{5.45}$$

where $\mathbf{Q}_{ij}$ indicates the control points of the patch and w_{ij} indicates their weights. This equation expresses the bi-7th degree rational Bézier surface.

A bicubic Gregory patch can be transformed through these steps into bi-7th degree rational Bézier surface. Figure 5.24 illustrates the control points of the bicubic Gregory patch and those of the bi-7th degree rational Bézier surface.

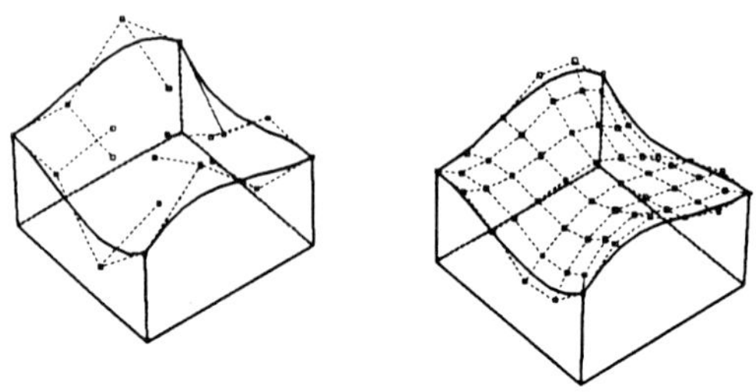

Fig. 5.24. Transformation of a Gregory patch to a rational Bézier surface

The same procedure can be applied when transforming a rational boundary Gregory patch to a rational Bézier patch. However, according to Equation (5.14), the multiple of the denominator would be different from $W(u,v)$. So, the multiple is set to $T(u,v)$ as follows:

$$T(u,v) = (u^2+v^2)((1-u)^2+v^2)(u^2+(1-v)^2)((1-u)^2+(1-v)^2) \tag{5.46}$$

Equation (5.46) is an 8th degree equation about u and v. It can be transformed into the equation of an 11th degree rational Bézier surface, by applying the same method as that used to transform a Gregory patch into a rational boundary Gregory patch, which would give:

$$\mathbf{S}(u,v) = \frac{\sum_{i=0}^{11}\sum_{j=0}^{11} B_i^{11}(u)B_j^{11}(v)w_{ij}\mathbf{Q}_{ij}}{\sum_{i=0}^{11}\sum_{j=0}^{11} B_i^{11}(u)B_j^{11}(v)w_{ij}} \tag{5.47}$$

where $\mathbf{Q}_{ij}$ indicate the control points and w_{ij} indicate their weights ($i = 0, \cdots, 11; j = 0, \cdots, 11$).

As explained, the Gregory patch or the rational boundary Gregory patch is the special case of the rational Bézier surface, and using the method previously discussed, the surface can be transformed into the NURBS.

5.13 Programs

This section shows two programs: "surfPosition", the program to obtain the coordinates of a point on a surface by using the surface data as outlined in Subsection 3.4.4 and the parameter, and "surfUDeriv", the program to obtain the partial derivative vector in the u direction of a surface by using the surface data and the parameters.

In DESIGNBASE, the surface data is defined by the following structure:

```
struct dbSurface {
    int     type;               /* Type of a surface */
    union {                     /* Surface data */
      struct dbPlane                      pl; /* Plane */
      struct dbSphere                     sp; /* Sphere */
      struct dbCylinder                   cy; /* Cylinder */
      struct dbCone                       co; /* Cone */
      struct dbBezier_patch               bp; /* Bezier patch */
      struct dbRational_Bezier_patch      rbp;/* Rational Bezier patch */
      struct dbGregory_patch              gp; /* Gregory patch */
      struct dbRational_Gregory_patch     rgp;/*
                                               * Rational boundary
                                               * Gregory patch
                                               */
    } *data;
};
struct dbBezier_patch {
    int udeg, vdeg;                 /* Degree in u and v direction */
    struct dbPoint  *ctlp;          /* Control points */
};
struct dbRational_Bezier_patch {
    int udeg, vdeg;                 /* Degree in u and v direction */
    struct dbHPoint *ctlp;          /* Control points with weight */
};
struct dbGregory_patch {
    int udeg, vdeg;                 /* Degree in u and v direction */
    struct dbPoint  *ctlp;          /* Control points */
};
struct dbRational_Gregory_patch {
    int udeg, vdeg;                 /* Degree in u and v direction */
    struct dbHPoint *ctlp;          /* Control points with weight */
};
```

The location of the control points on a bicubic Bézier or rational Bézier patch are as shown in Figure 5.25.

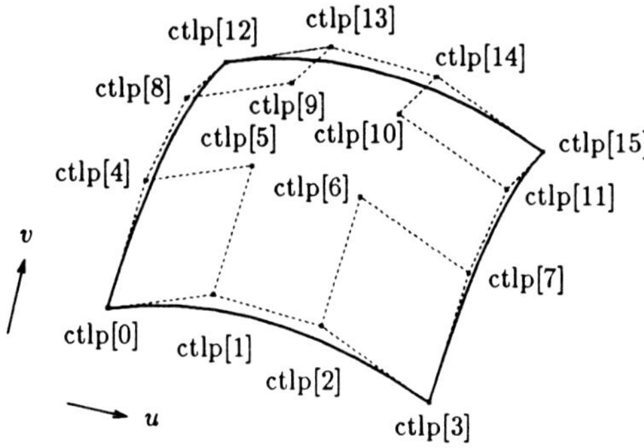

Fig. 5.25. Control points of a bicubic Bézier or rational Bézier patch

The control points of a bicubic Gregory patch are located as shown below in Figure 5.26.

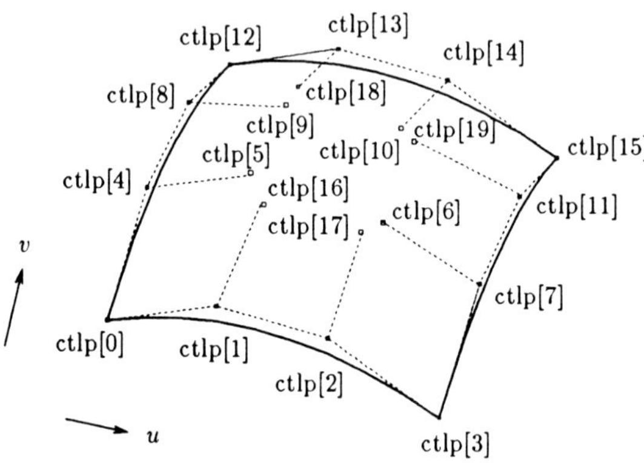

Fig. 5.26. Control points of bicubic Gregory patch

The control points of a bicubic rational boundary Gregory patch are located as shown below in Figure 5.27.

5.13.1 Program to obtain coordinates on a surface

The program "surfPosition" is to obtain the coordinates of a point on a surface using the input pointer to the surface data and the parameters. Please note that this program cannot be applied to plane and quadric surfaces since they are not parametric patches.

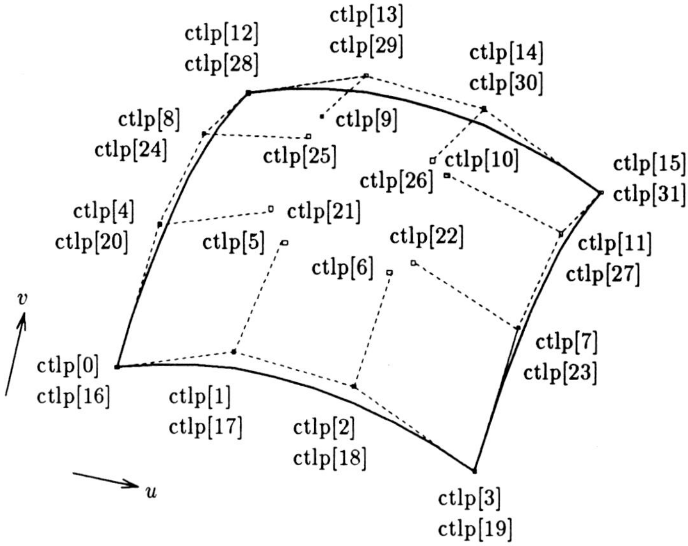

Fig. 5.27. Control points of bicubic rational boundary Gregory patch

```
/*
 * surfPosition --- Obtain coordinates of a point on a surface.
 *
 *  Parameters:
 *    sp       (I)  Surface patch
 *    u, v     (I)  Parameters
 *    pnt      (O)  Point on the patch
 */
void
surfPosition(sp, u, v, pnt)
    struct dbSurface *sp;
    double u, v;
    struct dbPoint *pnt;
{
    switch (sp->type) {
    case DB_BEZPATCH:
        surfBezPosition(&sp->data->bp, u, v, pnt);
        return;
    case DB_GREGPATCH:
        surfGrPosition(&sp->data->gp, u, v, pnt);
        return;
    case DB_RBEZPATCH:
        surfRBezPosition(&sp->data->rbp, u, v, pnt);
        return;
    case DB_RGREGPATCH:
```

```
        surfRGrPosition(&sp->data->rgp, u, v, pnt);
        return;
    default:
        fprintf(stderr, "illegal surface type\n");
        return;
    }
}

/*
 * surfBezPosition --- Obtain coordinates of a point
 *                     on a Bezier patch.
 *
 *  Parameters:
 *    bp      (I)  Bezier patch
 *    u, v    (I)  Parameters
 *    pnt     (O)  Point on the patch
 */
void
surfBezPosition(bp, u, v, pnt)
    struct dbBezier_patch *bp;
    double u, v;
    struct dbPoint *pnt;
{
    double u1 = 1.0 - u;
    double v1 = 1.0 - v;
    double *culist = (double *) 0;
    double *cvlist = (double *) 0;
    double *ulist = (double *) 0;
    double *u1list = (double *) 0;
    double *vlist = (double *) 0;
    double *v1list = (double *) 0;
    register int i, j;
    double vtmp, tmp;

    /*
     * Allocate the necessary area.
     */
    culist = (double *) malloc((bp->udeg+1) * sizeof(double));
    cvlist = (double *) malloc((bp->vdeg+1) * sizeof(double));
    ulist = (double *) malloc((bp->udeg+1) * sizeof(double));
    u1list = (double *) malloc((bp->udeg+1) * sizeof(double));
    vlist = (double *) malloc((bp->vdeg+1) * sizeof(double));
    v1list = (double *) malloc((bp->vdeg+1) * sizeof(double));
    if (culist == 0 || cvlist == 0 || ulist == 0 || u1list == 0 ||
        vlist == 0 || v1list == 0) {
            fprintf(stderr, "malloc error\n");
            exit(1);
    }
```

```
    /*
     * See Section 4.7 for "GetCombList" and "GetPowerList".
     */
    GetCombList(bp->udeg, culist);
    GetCombList(bp->vdeg, cvlist);
    GetPowerList(u, bp->udeg, ulist);
    GetPowerList(u1, bp->udeg, u1list);
    GetPowerList(v, bp->vdeg, vlist);
    GetPowerList(v1, bp->vdeg, v1list);

#define BP(i,j)         (bp->ctlp + (bp->udeg+1)*(i) + (j))
    pnt->x = pnt->y = pnt->z = 0.0;
    for (i = 0; i <= bp->vdeg; i++) {
        vtmp = cvlist[i] * vlist[i] * v1list[bp->vdeg-i];
        for (j = 0; j <= bp->udeg; j++) {
        tmp = vtmp * culist[j] * ulist[j] * u1list[bp->udeg-j];
            pnt->x += tmp * ((*BP(i,j)).x);
            pnt->y += tmp * ((*BP(i,j)).y);
            pnt->z += tmp * ((*BP(i,j)).z);
        }
    }
#undef BP
    free(culist); free(cvlist);
    free(ulist); free(u1list);
    free(vlist); free(v1list);
}

/*
 * surfGrPosition --- Obtain coordinates of a point
 *                    on a Gregory patch.
 *
 *  Parameters:
 *    gp      (I)  Gregory patch
 *    u, v    (I)  Parameters
 *    pnt     (O)  Point on the patch
 */
void
surfGrPosition(gp, u, v, pnt)
    struct dbGregory_patch *gp;
    double u, v;
    struct dbPoint *pnt;
{
    struct dbBezier_patch b;

    /*
     * Set bicubic Bezier patch index.
     */
```

```
    b.udeg = gp->udeg; b.vdeg = gp->vdeg;
    b.ctlp = (struct dbPoint *)
        malloc((gp->udeg+1)*(gp->vdeg+1)*sizeof(struct dbPoint));
    if (b.ctlp == 0) {
        fprintf(stderr, "malloc error\n");
        exit(1);
    }

    surfSetGrPnt(gp, &b, u, v);
    surfBezPosition(&b, u, v, pnt);
    free(b.ctlp);
}

/*
 * surfRBezPosition --- Obtain coordinates of a point
 *                      on a rational Bezier patch.
 *
 *  Parameters:
 *    bp      (I)  Rational Bezier patch
 *    u, v    (I)  Parameters
 *    pnt     (O)  Point on the patch
 */
void
surfRBezPosition(bp, u, v, pnt)
    struct dbRational_Bezier_patch *bp;
    double u, v;
    struct dbPoint *pnt;
{
    double u1 = 1.0 - u;
    double v1 = 1.0 - v;
    double *culist = (double *) 0;
    double *cvlist = (double *) 0;
    double *ulist = (double *) 0;
    double *u1list = (double *) 0;
    double *vlist = (double *) 0;
    double *v1list = (double *) 0;
    double w;
    register int i, j;
    double tmp, vtmp;

    culist = (double *) malloc((bp->udeg+1) * sizeof(double));
    cvlist = (double *) malloc((bp->vdeg+1) * sizeof(double));
    ulist = (double *) malloc((bp->udeg+1) * sizeof(double));
    u1list = (double *) malloc((bp->udeg+1) * sizeof(double));
    vlist = (double *) malloc((bp->vdeg+1) * sizeof(double));
    v1list = (double *) malloc((bp->vdeg+1) * sizeof(double));
    if (culist == 0 || cvlist == 0 || ulist == 0 || u1list == 0 ||
        vlist == 0 || v1list == 0) {
```

```
            fprintf(stderr, "malloc error\n");
            exit(1);
    }

    /*
     * See Section 4.7 for "GetCombList" and "GetPowerList".
     */
    GetCombList(bp->udeg, culist);
    GetCombList(bp->vdeg, cvlist);
    GetPowerList(u, bp->udeg, ulist);
    GetPowerList(u1, bp->udeg, u1list);
    GetPowerList(v, bp->vdeg, vlist);
    GetPowerList(v1, bp->vdeg, v1list);

#define BP(i,j)         (bp->ctlp + (bp->udeg+1)*(i) + (j))
    w = 0.0;
    pnt->x = pnt->y = pnt->z = 0.0;
    for (i = 0; i <= bp->vdeg; i++) {
        vtmp = cvlist[i] * vlist[i] * v1list[bp->vdeg-i];
        for (j = 0; j <= bp->udeg; j++) {
            tmp = vtmp * culist[j] * ulist[j] * u1list[bp->udeg-j]
                * ((*BP(i,j)).w);
            pnt->x += tmp * ((*BP(i,j)).x);
            pnt->y += tmp * ((*BP(i,j)).y);
            pnt->z += tmp * ((*BP(i,j)).z);
            w += tmp;
        }
    }
#undef BP

    if (fabs(w) < EPS) {
        pnt->x = pnt->y = pnt->z = HUGE;
    } else {
        pnt->x /= w; pnt->y /= w; pnt->z /= w;
    }

    free(culist); free(cvlist);
    free(ulist); free(u1list);
    free(vlist); free(v1list);
}

/*
 * surfRGrPosition --- Obtain coordinates of a point
 *                     on a rational boundary Gregory patch.
 *
 *  Parameters:
 *    rgp     (I)  Rational boundary Gregory patch
 *    u, v    (I)  Parameters
```

```
 *    pnt     (O)  Point on the patch
 */
void
surfRGr2Position(rgp, u, v, pnt)
    struct dbRational_Gregory_patch *rgp;
    double u, v;
    struct dbPoint *pnt;
{
    register int i, j;
    struct dbRational_Bezier_patch b;
    double u1 = 1.0 - u;
    double v1 = 1.0 - v;
    double *ulist, *u1list, *vlist, *v1list;
    double *culist, *cvlist;
    double w, tmp, vtmp;

    b.udeg = rgp->udeg; b.vdeg = rgp->vdeg;
    b.ctlp = (struct dbHPoint *)
        malloc((rgp->udeg+1)*(rgp->vdeg+1)*sizeof(struct dbHPoint));
    if (b.ctlp == 0) {
        fprintf(stderr, "malloc error\n");
        exit(1);
    }

    culist = (double *) malloc((rgp->udeg+1) * sizeof(double));
    cvlist = (double *) malloc((rgp->vdeg+1) * sizeof(double));
    ulist = (double *) malloc((rgp->udeg+1) * sizeof(double));
    u1list = (double *) malloc((rgp->udeg+1) * sizeof(double));
    vlist = (double *) malloc((rgp->vdeg+1) * sizeof(double));
    v1list = (double *) malloc((rgp->vdeg+1) * sizeof(double));
    if (culist == 0 || cvlist == 0 || ulist == 0 || u1list == 0 ||
        vlist == 0 || v1list == 0) {
            fprintf(stderr, "malloc error\n");
            exit(1);
    }

    surfSetRGrPnt(rgp, &b, u, v);

    /*
     * Check u and v whether they are 0 or 1.
     */
    if (surfEndParameter(u, v)) {
        sfRBezPosition(&b, u, v, pnt);
        goto EXIT;
    }

    /*
     * See Section 4.7 for "GetCombList" and "GetPowerList".
```

```
     */
    GetCombList(rgp->udeg, culist);
    GetCombList(rgp->vdeg, cvlist);
    GetPowerList(u, rgp->udeg, ulist);
    GetPowerList(u1, rgp->udeg, u1list);
    GetPowerList(v, rgp->vdeg, vlist);
    GetPowerList(v1, rgp->vdeg, v1list);

#define RGP(i, j)       (b.ctlp + (rgp->udeg+1)*(i) + (j))
    w = 0.0;
    pnt->x = pnt->y = pnt->z = 0.0;
    for (i = 0; i <= rgp->vdeg; i++) {
        vtmp = cvlist[i] * vlist[i] * v1list[rgp->vdeg-i];
        for (j = 0; j <= rgp->udeg; j++) {
            tmp = vtmp * culist[j] * ulist[j]*u1list[rgp->udeg-j];
            pnt->x += tmp * ((*RGP(i,j)).x);
            pnt->y += tmp * ((*RGP(i,j)).y);
            pnt->z += tmp * ((*RGP(i,j)).z);
            w += tmp * ((*RGP(i,j)).w);
        }
    }
#undef RGP

    if (fabs(w) < EPS) {
        pnt->x = pnt->y = pnt->z = HUGE;
    } else {
        pnt->x /= w; pnt->y /= w; pnt->z /= w;
    }

    free(b.ctlp);
    free(culist); free(cvlist);
    free(ulist); free(u1list);
    free(vlist); free(v1list);
}

/*
 * surfSetGrPnt --- Obtain coordinates of interpolating control
 *                  points of a Gregory patch.
 *
 *  Parameters:
 *    gp      (I)  Gregory patch
 *    bp      (O)  Interpolating Bezier patch
 *    u, v    (I)  Parameter
 */
void
surfSetGrPnt(gp, bp, u, v)
    struct dbGregory_patch *gp;
    struct dbBezier_patch *bp;
```

```
    double u, v;
{
    double _u = 1.0 - u;
    double _v = 1.0 - v;
    register struct dbPoint *p0, *p1, *p2, *p3;
    register int ind;

    /*
     * First, copy (gp->udeg+1)*(gp->vdeg+1) control points
     * of the Gregory patch.
     */
    bcopy((char *) gp->ctlp, (char *) bp->ctlp,
        (gp->udeg+1)*(gp->vdeg+1)*sizeof(struct dbPoint));

    /*
     * Check whether the specified parameter is on corner or not.
     */
    if (surfEndParameter(u, v))
        return;

    /*
     * Set the inner control points.
     */
    p0 = bp->ctlp + gp->udeg + 2;
    p1 = bp->ctlp + (gp->udeg + 1) * 2 - 2;
    p2 = bp->ctlp + (gp->udeg + 1) * (gp->vdeg - 1) + 1;
    p3 = bp->ctlp + (gp->udeg + 1) * (gp->vdeg) - 2;

    ind = (gp->udeg+1)*(gp->vdeg+1);

    p0->x *= v; p0->y *= v; p0->z *= v;
    p0->x += u*gp->ctlp[ind].x;
    p0->y += u*gp->ctlp[ind].y;
    p0->z += u*gp->ctlp[ind++].z;
    p0->x /= (u+v); p0->y /= (u+v); p0->z /= (u+v);

    p1->x *= v; p1->y *= v; p1->z *= v;
    p1->x += _u*gp->ctlp[ind].x;
    p1->y += _u*gp->ctlp[ind].y;
    p1->z += _u*gp->ctlp[ind++].z;
    p1->x /= (_u+v); p1->y /= (_u+v); p1->z /= (_u+v);

    p2->x *= _v; p2->y *= _v; p2->z *= _v;
    p2->x += u*gp->ctlp[ind].x;
    p2->y += u*gp->ctlp[ind].y;
    p2->z += u*gp->ctlp[ind++].z;
    p2->x /= (u+_v); p2->y /= (u+_v); p2->z /= (u+_v);
```

```
    p3->x *= _v; p3->y *= _v; p3->z *= _v;
    p3->x += _u*gp->ctlp[ind].x;
    p3->y += _u*gp->ctlp[ind].y;
    p3->z += _u*gp->ctlp[ind++].z;
    p3->x /= (_u+_v); p3->y /= (_u+_v); p3->z /= (_u+_v);
}

/*
 * surfSetRGrPnt --- Obtain coordinates of interpolating control
 *                   points of a bicubic rational boundary
 *                   Gregory patch.
 *
 *  Parameters:
 *    rgp     (I)  Bicubic rational boundary Gregory patch
 *    bp      (O)  Interpolating rational Bezier patch
 *    u, v    (I)  Parameters
 */
void
surfSetRGrPnt(rgp, bp, u, v)
    struct dbRational_Gregory_patch *rgp;
    struct dbRational_Bezier_patch *bp;
    double u, v;
{
    int size;
    register int i, j;
    double uu, vv, __uu, __vv;
    register struct dbHPoint *p0, *p1;
    register struct dbHPoint *q;

    if(fabs(u) < EPS) {
        if(fabs(v) < EPS || fabs(v-1.0) < EPS) {
            size = (rgp->udeg+1)*(rgp->vdeg+1);
            bcopy((char *) rgp->ctlp, (char *) bp->ctlp,
                                size*sizeof(struct dbHPoint));
            return;
        }
    } else if(fabs(u - 1.0) < EPS) {
        if(fabs(v) < EPS || fabs(v - 1.0) < EPS) {
            size = (rgp->udeg+1)*(rgp->vdeg+1);
            bcopy((char*)&rgp->ctlp[(rgp->udeg+1)*(rgp->vdeg+1)],
                (char *)bp->ctlp, size*sizeof(struct dbHPoint));
            return;
        }
    }

    p0 = &rgp->ctlp[(rgp->udeg+1)*(rgp->vdeg+1)];
    p1 = rgp->ctlp;
    q = bp->ctlp;
```

```
    uu = u * u; vv = v * v;
    __uu = (1.0-u) * (1.0-u); __vv = (1.0-v) * (1.0-v);
    for (i = 0; i < 4; i++) {
        for (j = 0; j < 4; j++) {
            if (i <= 1 && j <= 1) {
                q->x = (uu * p0->x*p0->w + vv * p1->x*p1->w)
                    / (uu + vv);
                q->y = (uu * p0->y*p0->w + vv * p1->y*p1->w)
                    / (uu + vv);
                q->z = (uu * p0->z*p0->w + vv * p1->z*p1->w)
                    / (uu + vv);
                q->w = (uu * p0->w + vv * p1->w) / (uu + vv);
            } else if (i <= 1 && j >= 2) {
                q->x = (__uu * p0->x*p0->w + vv * p1->x*p1->w)
                    / (__uu + vv);
                q->y = (__uu * p0->y*p0->w + vv * p1->y*p1->w)
                    / (__uu + vv);
                q->z = (__uu * p0->z*p0->w + vv * p1->z*p1->w)
                    / (__uu + vv);
                q->w = (__uu * p0->w + vv * p1->w)/(__uu + vv);
            } else if (i >= 2 && j <= 1) {
                q->x = (uu * p0->x*p0->w + __vv * p1->x*p1->w)
                    / (uu + __vv);
                q->y = (uu * p0->y*p0->w + __vv * p1->y*p1->w)
                    / (uu + __vv);
                q->z = (uu * p0->z*p0->w + __vv * p1->z*p1->w)
                    / (uu + __vv);
                q->w = (uu * p0->w + __vv * p1->w)/(uu + __vv);
            } else {
                ; assert(i >= 2 && j >= 2);
                q->x = (__uu * p0->x*p0->w + __vv * p1->x*p1->w)
                    / (__uu + __vv);
                q->y = (__uu * p0->y*p0->w + __vv * p1->y*p1->w)
                    / (__uu + __vv);
                q->z = (__uu * p0->z*p0->w + __vv * p1->z*p1->w)
                    / (__uu + __vv);
                q->w = (__uu * p0->w + __vv * p1->w)
                    / (__uu + __vv);
            }
            p0++; p1++; q++;
        }
    }
}

/*
 * surfEndParameter --- Check that parameter specifies corner.
 *
```

```
 *  Parameters:
 *    u, v    (I)  Parameter
 *  Returns:
 *    1   corner
 *    0   nor corner
 */
int
surfEndParameter(u, v)
    double u, v;
{
    if (fabs(u) < EPS && fabs(u) < EPS)
        return 1;
    if (fabs(u) < EPS && fabs(1.0 - u) < EPS)
        return 1;
    if (fabs(1.0 - u) < EPS && fabs(u) < EPS)
        return 1;
    if (fabs(1.0 - u) < EPS && fabs(1.0 - u) < EPS)
        return 1;
    return 0;
}
```

5.13.2 Program to obtain a partial derivative vector of a surface

The program "surfUDeriv" is to obtain a partial derivative vector of a surface in the u direction using the input pointer to the surface data and the parameters.

```
/*
 * surfUDeriv --- Obtain a partial derivative vector on a surface
 *                in u direction.
 *
 *  Parameters:
 *    sp      (I)  Surface patch
 *    u, v    (I)  Parameters
 *    drv     (O)  Partial derivative vector
 */
void
surfUDeriv(sp, u, v, drv)
    struct dbSurface *sp;
    double u, v;
    struct dbVector *drv;
{
    switch (sp->type) {
    case DB_BEZPATCH:
        surfBezUDeriv(&sp->data->bp, u, v, drv);
        return;
    case DB_GREGPATCH:
        surfGrUDeriv(&sp->data->gp, u, v, drv);
```

```
        return;
    case DB_RBEZPATCH:
        surfRBezUDeriv(&sp->data->rbp, u, v, drv);
        return;
    case DB_RGREGPATCH:
        surfRGrUDeriv(&sp->data->rgp, u, v, drv);
        return;
    default:
        fprintf(stderr, "illegal surface type is specified\n");
        return;
    }
}

/*
 * surfBezUDeriv --- Obtain a partial derivative vector
 *                   on a Bezier patch in u direction.
 *
 *  Parameters:
 *    bp      (I)  Bezier patch
 *    u, v    (I)  Parameters
 *    drv     (O)  Partial derivative vector
 */
void
surfBezUDeriv(bp, u, v, drv)
    struct dbBezier_patch *bp;
    double u, v;
    struct dbVector *drv;
{
    register int i, j;
    register double *culist = (double *) 0;
    register double *cvlist = (double *) 0;
    register double *ulist = (double *) 0;
    register double *u1list = (double *) 0;
    register double *vlist = (double *) 0;
    register double *v1list = (double *) 0;
    double u1 = 1.0 - u;
    double v1 = 1.0 - v;
    double vtmp, tmp;

    culist = (double *) malloc((bp->udeg) * sizeof(double));
    cvlist = (double *) malloc((bp->vdeg+1) * sizeof(double));
    ulist = (double *) malloc((bp->udeg) * sizeof(double));
    u1list = (double *) malloc((bp->udeg) * sizeof(double));
    vlist = (double *) malloc((bp->vdeg+1) * sizeof(double));
    v1list = (double *) malloc((bp->vdeg+1) * sizeof(double));
    if (culist == 0 || cvlist == 0 || ulist == 0 || u1list == 0 ||
        vlist == 0 || v1list == 0) {
            fprintf(stderr, "malloc error\n");
```

```
            exit(1);
    }

    /*
     * See Section 4.7 for "GetCombList" and "GetPowerList".
     */
    GetCombList(bp->udeg - 1, culist);
    GetCombList(bp->vdeg, cvlist);
    GetPowerList(u, bp->udeg - 1, ulist);
    GetPowerList(u1, bp->udeg - 1, u1list);
    GetPowerList(v, bp->vdeg, vlist);
    GetPowerList(v1, bp->vdeg, v1list);

#define BP(i,j)         (bp->ctlp + (bp->udeg+1)*(i) + (j))
    drv->x = drv->y = drv->z = 0.0;
    for (i = 0; i <= bp->vdeg; i++) {
        vtmp = cvlist[i] * vlist[i] * v1list[bp->vdeg-i];
        for (j = 0; j < bp->udeg; j++) {
            tmp = vtmp * culist[j]*ulist[j]*u1list[bp->udeg-j-1];
            drv->x += tmp * (((*BP(i,j+1)).x) - ((*BP(i,j)).x));
            drv->y += tmp * (((*BP(i,j+1)).y) - ((*BP(i,j)).y));
            drv->z += tmp * (((*BP(i,j+1)).z) - ((*BP(i,j)).z));
        }
    }
#undef BP
    drv->x *= (double) (bp->udeg);
    drv->y *= (double) (bp->udeg);
    drv->z *= (double) (bp->udeg);

    free(culist); free(cvlist);
    free(ulist); free(u1list);
    free(vlist); free(v1list);
}

/*
 * surfGrUDeriv --- Obtain a partial derivative vector
 *                  on a Gregory patch in u direction.
 *
 *  Parameters:
 *    gp      (I)  Gregory patch
 *    u, v    (I)  Parameter
 *    drv     (O)  Partial derivative vector
 */
void
surfGrUDeriv(gp, u, v, drv)
    struct dbGregory_patch *gp;
    double u, v;
    struct dbVector *drv;
```

```
{
    struct dbBezier_patch b;
    struct dbVector vec;
    double uv ,_uv, u_v, _u_v, _v;
    register struct dbPoint *p0, *p1, *p2, *p3;
    register int ind;
    register int ind0, ind1, ind2, ind3;

    b.udeg = gp->udeg; b.vdeg = gp->vdeg;
    b.ctlp = (struct dbPoint *)
        malloc((gp->udeg+1)*(gp->vdeg+1)*sizeof(struct dbPoint));
    if (b.ctlp == 0) {
        fprintf(stderr, "malloc error\n");
        exit(1);
    }
    if (surfEndParameter(u, v)) {
        bcopy(gp->ctlp, b.ctlp,
                (b.udeg+1)*(b.vdeg+1)*sizeof(struct dbPoint));
        surfBezUDeriv(&b, u, v, drv);
        free(b.ctlp);
        return;
    } else {
        surfSetGrPnt(gp, &b, u, v);
        surfBezUDeriv(&b, u, v, &vec);
    }
    bzero((char *) b.ctlp,
                (b.udeg+1)*(b.vdeg+1)*sizeof(struct dbPoint));
    uv = u + v;
    _uv = 1.0 - u + v;
    u_v = u + (_v = 1.0 - v);
    _u_v = 2.0 - uv;

    ind0 = gp->udeg + 2; p0 = b.ctlp + ind0;
    ind1 = (gp->udeg + 1) * 2 - 2; p1 = b.ctlp + ind1;
    ind2 = (gp->udeg + 1)*(gp->vdeg - 1) + 1; p2 = b.ctlp + ind2;
    ind3 = (gp->udeg + 1) * (gp->vdeg) - 2; p3 = b.ctlp + ind3;
    ind = (gp->udeg+1)*(gp->vdeg+1);

    p0->x = v*(gp->ctlp[ind].x - gp->ctlp[ind0].x)/(uv * uv);
    p0->y = v*(gp->ctlp[ind].y - gp->ctlp[ind0].y)/(uv * uv);
    p0->z = v*(gp->ctlp[ind++].z - gp->ctlp[ind0].z)/(uv * uv);
    p1->x = -v*(gp->ctlp[ind].x - gp->ctlp[ind1].x)/(_uv * _uv);
    p1->y = -v*(gp->ctlp[ind].y - gp->ctlp[ind1].y)/(_uv * _uv);
    p1->z = -v*(gp->ctlp[ind++].z - gp->ctlp[ind1].z)/(_uv * _uv);
    p2->x = _v*(gp->ctlp[ind].x - gp->ctlp[ind2].x)/(u_v * u_v);
    p2->y = _v*(gp->ctlp[ind].y - gp->ctlp[ind2].y)/(u_v * u_v);
    p2->z = _v*(gp->ctlp[ind++].z - gp->ctlp[ind2].z)/(u_v * u_v);
    p3->x = -_v*(gp->ctlp[ind].x - gp->ctlp[ind3].x)/(_u_v * _u_v);
```

```
    p3->y = -_v*(gp->ctlp[ind].y - gp->ctlp[ind3].y)/(_u_v * _u_v);
    p3->z = -_v*(gp->ctlp[ind++].z - gp->ctlp[ind3].z)/(_u_v * _u_v);
    surfBezPosition(&b, u, v, drv);

    drv->x += vec.x; drv->y += vec.y; drv->z += vec.z;

    free(b.ctlp);
}

/*
 * surfRBezUDeriv --- Obtain a partial derivative vector
 *                    on a rational Bezier patch in u direction.
 *
 *  Parameters:
 *    rbp     (I)  Rational Bezier patch
 *    u, v    (I)  Parameter
 *    drv     (O)  Partial derivative vector
 */
void
surfRBezUDeriv(rbp, u, v, drv)
    struct dbRational_Bezier_patch *rbp;
    double u, v;
    struct dbVector *drv;
{
    struct dbPoint pdrv;
    double wdrv;
    register int i, j;
    register double *culist = (double *) 0;
    register double *cvlist = (double *) 0;
    register double *_culist = (double *) 0;
    register double *ulist = (double *) 0;
    register double *u1list = (double *) 0;
    register double *vlist = (double *) 0;
    register double *v1list = (double *) 0;
    struct dbPoint p;
    double w;
    double u1 = 1.0 - u;
    double v1 = 1.0 - v;
    double vtmp, tmp;

    culist = (double *) malloc((rbp->udeg+1) * sizeof(double));
    cvlist = (double *) malloc((rbp->vdeg+1) * sizeof(double));
    _culist = (double *) malloc((rbp->udeg) * sizeof(double));
    ulist = (double *) malloc((rbp->udeg+1) * sizeof(double));
    u1list = (double *) malloc((rbp->udeg+1) * sizeof(double));
    vlist = (double *) malloc((rbp->vdeg+1) * sizeof(double));
    v1list = (double *) malloc((rbp->vdeg+1) * sizeof(double));
    if (culist == 0 || _culist == 0 || cvlist == 0 || ulist == 0 ||
```

```
        u1list == 0 || vlist == 0 || v1list == 0) {
            fprintf(stderr, "malloc error\n");
            exit(1);
    }

    /*
     * See Section 4.7 for "GetCombList" and "GetPowerList".
     */
    GetCombList(rbp->udeg - 1, _culist);
    GetCombList(rbp->vdeg, cvlist);
    for (i = 0; i < rbp->udeg - 1; i++) {
        culist[i+1] = _culist[i] + _culist[i+1];
    }
    culist[0] = culist[rbp->udeg] = 1;

    GetPowerList(u, rbp->udeg, ulist);
    GetPowerList(u1, rbp->udeg, u1list);
    GetPowerList(v, rbp->vdeg, vlist);
    GetPowerList(v1, rbp->vdeg, v1list);

#define BP(i,j)         (rbp->ctlp + (rbp->udeg+1)*(i) + (j))
    w = 0.0;
    p.x = p.y = p.z = 0.0;
    for (i = 0; i <= rbp->vdeg; i++) {
        vtmp = cvlist[i] * vlist[i] * v1list[rbp->vdeg-i];
        for (j = 0; j <= rbp->udeg; j++) {
            tmp = vtmp * culist[j] * ulist[j] * u1list[rbp->udeg-j]
                * ((*BP(i,j)).w);
            p.x += tmp * ((*BP(i,j)).x);
            p.y += tmp * ((*BP(i,j)).y);
            p.z += tmp * ((*BP(i,j)).z);
            w += tmp;
        }
    }
    wdrv = 0.0;
    pdrv.x = pdrv.y = pdrv.z = 0.0;
    for (i = 0; i <= rbp->vdeg; i++) {
        vtmp = cvlist[i] * vlist[i] * v1list[rbp->vdeg-i];
        for (j = 0; j < rbp->udeg; j++) {
            tmp = vtmp * _culist[j]*ulist[j]*u1list[rbp->udeg-1-j];
            pdrv.x += tmp * (((*BP(i,j+1)).w)*((*BP(i,j+1)).x) -
                             ((*BP(i,j)).w)*((*BP(i,j)).x));
            pdrv.y += tmp * (((*BP(i,j+1)).w)*((*BP(i,j+1)).y) -
                             ((*BP(i,j)).w)*((*BP(i,j)).y));
            pdrv.z += tmp * (((*BP(i,j+1)).w)*((*BP(i,j+1)).z) -
                             ((*BP(i,j)).w)*((*BP(i,j)).z));
            wdrv += tmp * ((*BP(i,j+1)).w - (*BP(i,j)).w);
        }
```

```
    }
    pdrv.x *= (double) rbp->udeg;
    pdrv.y *= (double) rbp->udeg;
    pdrv.z *= (double) rbp->udeg;
    wdrv *= (double) rbp->udeg;
#undef BP

    if (fabs(w) < EPS) {
        drv->x = drv->y = drv->z = HUGE;
    } else {
        drv->x = (pdrv.x * w - p.x * wdrv) / (w * w);
        drv->y = (pdrv.y * w - p.y * wdrv) / (w * w);
        drv->z = (pdrv.z * w - p.z * wdrv) / (w * w);
    }

    free(culist); free(cvlist); free(_culist);
    free(ulist); free(u1list);
    free(vlist); free(v1list);
}

/*
 * surfRGrUDeriv --- Obtain a partial derivative vector on a rational
 *                   boundary Gregory patch in u direction.
 *
 *  Parameters:
 *    rgp     (I)  Rational boundary Gregory patch
 *    u, v    (I)  Parameter
 *    drv     (O)  Partial derivative vector
 */
void
surfRGrUDeriv(rgp, u, v, drv)
    struct dbRational_Gregory_patch *rgp;
    double u, v;
    struct dbVector *drv;
{
    register int i, j;
    register double *culist, *cvlist, *_culist;
    register double *ulist, *u1list, *vlist, *v1list;
    double u1 = 1.0 - u;
    double v1 = 1.0 - v;
    struct dbPoint p
    struct dbVector pdrv;
    struct dbHPoint pnt, dpnt, pnt1;
    double w, wdrv;
    double tmp, vtmp;

    /*
     * Limit the degree of the rational boundary Gregory patch
```

```
 * to bicubic.
 */
culist = (double *) malloc((rgp->udeg+1) * sizeof(double));
cvlist = (double *) malloc((rgp->vdeg+1) * sizeof(double));
_culist = (double *) malloc(rgp->udeg * sizeof(double));
ulist = (double *) malloc((rgp->udeg+1) * sizeof(double));
u1list = (double *) malloc((rgp->udeg+1) * sizeof(double));
vlist = (double *) malloc((rgp->vdeg+1) * sizeof(double));
v1list = (double *) malloc((rgp->vdeg+1) * sizeof(double));
if (culist == 0 || _culist == 0 || cvlist == 0 || ulist == 0 ||
    u1list == 0 || vlist == 0 || v1list == 0) {
        fprintf(stderr, "malloc error\n");
        exit(1);
}

/*
 * See Section 4.7 for "GetCombList" and "GetPowerList".
 */
GetCombList(rgp->udeg - 1, _culist);
GetCombList(rgp->vdeg, cvlist);
for (i = 0; i < rgp->udeg - 1; i++) {
    culist[i+1] = _culist[i] + _culist[i+1];
}
culist[0] = culist[rgp->udeg] = 1;
GetPowerList(u, rgp->udeg, ulist);
GetPowerList(u1, rgp->udeg, u1list);
GetPowerList(v, rgp->vdeg, vlist);
GetPowerList(v1, rgp->vdeg, v1list);

if (surfEndParameter(u, v)) {
    struct dbRational_Bezier_patch b;

    b.udeg = rgp->udeg;
    b.vdeg = rgp->vdeg;
    b.ctlp = (struct dbHPoint *)
        malloc((b.udeg+1)*(b.vdeg+1)*sizeof(struct dbHPoint));
    if (b.ctlp == 0) {
        fprintf(stderr, "malloc error\n");
        exit(1);
    }
    bcopy(rgp->ctlp, b.ctlp,
            (b.udeg+1)*(b.vdeg+1)*sizeof(struct dbHPoint));
    surfRBezUDeriv(&b, u, v, drv);
    free(b.ctlp);
    goto EXIT;
}

p.x = p.y = p.z = w = 0.0;
```

```
    pdrv.x = pdrv.y = pdrv.z = wdrv = 0.0;
    for (i = 0; i <= rgp->vdeg; i++) {
        vtmp = cvlist[i]*vlist[i] * v1list[rgp->vdeg-i];
        for (j = 0; j <= rgp->udeg; j++) {
            RGREG2PNT(rgp, i, j, u, v, &pnt);
            RGREG2DU(rgp, i, j, u, v, &dpnt);
            tmp = vtmp * culist[j] * ulist[j]*u1list[rgp->udeg-j];
            p.x += tmp * pnt.x;
            p.y += tmp * pnt.y;
            p.z += tmp * pnt.z;
            w += tmp * pnt.w;
            pdrv.x += tmp * dpnt.x;
            pdrv.y += tmp * dpnt.y;
            pdrv.z += tmp * dpnt.z;
            wdrv += tmp * dpnt.w;
        }
    }
    for (i = 0; i <= rgp->vdeg; i++) {
        vtmp = cvlist[i] * vlist[i]*v1list[rgp->vdeg-i];
        for (j = 0; j < rgp->udeg; j++) {
            tmp = vtmp * rgp->udeg*_culist[j] *
                        ulist[j]*u1list[rgp->udeg-j-1];
            RGREG2PNT(rgp, i, j, u, v, &pnt);
            RGREG2PNT(rgp, i, j+1, u, v, &pnt1);
            pdrv.x += tmp * (pnt1.x - pnt.x);
            pdrv.y += tmp * (pnt1.y - pnt.y);
            pdrv.z += tmp * (pnt1.z - pnt.z);
            wdrv += tmp * (pnt1.w - pnt.w);
        }
    }

    if (fabs(w) < EPS) {
        drv->x = drv->y = drv->z = HUGE;
    } else {
        drv->x = (pdrv.x * w - p.x * wdrv) / (w * w);
        drv->y = (pdrv.y * w - p.y * wdrv) / (w * w);
        drv->z = (pdrv.z * w - p.z * wdrv) / (w * w);
    }

EXIT:
    free(culist); free(cvlist); free(_culist);
    free(ulist); free(u1list);
    free(vlist); free(v1list);
}

/*
 * RGREG2PNT --- Obtain interpolating control points of a bicubic
 *               rational boundary Gregory patch.
```

```
 *
 *  Parameters:
 *    rgp      (I)  Bicubic rational boundary Gregory patch
 *    i, j     (I)  Index
 *    u, v     (I)  Parameters
 *    pnt      (O)  Interpolating points
 */
void
RGREG2PNT(rgp, i, j, u, v, pnt)
    struct dbRational_Gregory_patch *rgp;
    int i, j;
    double u, v;
    struct dbHPoint *pnt;
{
    register struct dbHPoint *gp0, *gp1;
    double uu, vv, __uu, __vv;
    double tmp;

#define GP0(i,j)  (rgp->ctlp + (rgp->vdeg+1)*(rgp->udeg+1+i) + (j))
#define GP1(i,j)  (rgp->ctlp + (rgp->vdeg+1)*(i) + (j))
    /*
     * Limit the degree of the rational boundary Gregory patch
     * to bicubic.
     */
    gp0 = GP0(i,j);
    gp1 = GP1(i,j);
    if (i < 2) {
        vv = v * v;
        if (j < 2) {
            uu = u * u;
            tmp = uu + vv;
            pnt->x = (uu*gp0->x*gp0->w + vv*gp1->x*gp1->w)/tmp;
            pnt->y = (uu*gp0->y*gp0->w + vv*gp1->y*gp1->w)/tmp;
            pnt->z = (uu*gp0->z*gp0->w + vv*gp1->z*gp1->w)/tmp;
            pnt->w = (uu*gp0->w + vv*gp1->w)/tmp;
        } else {
            __uu = (1.0 - u) * (1.0 - u);
            tmp = __uu + vv;
            pnt->x = (__uu*gp0->x*gp0->w + vv*gp1->x*gp1->w)/tmp;
            pnt->y = (__uu*gp0->y*gp0->w + vv*gp1->y*gp1->w)/tmp;
            pnt->z = (__uu*gp0->z*gp0->w + vv*gp1->z*gp1->w)/tmp;
            pnt->w = (__uu*gp0->w+ vv*gp1->w)/tmp;
        }
    } else {
        __vv = (1.0 - v) * (1.0 - v);
        if (j < 2) {
            uu = u * u;
            tmp = uu + __vv;
```

```
            pnt->x = (uu*gp0->x*gp0->w + __vv*gp1->x*gp1->w)/tmp;
            pnt->y = (uu*gp0->y*gp0->w + __vv*gp1->y*gp1->w)/tmp;
            pnt->z = (uu*gp0->z*gp0->w + __vv*gp1->z*gp1->w)/tmp;
            pnt->w = (uu*gp0->w+ __vv*gp1->w)/tmp;
        } else {
            __uu = (1.0 - u) * (1.0 - u);
            tmp = __uu + __vv;
            pnt->x = (__uu*gp0->x*gp0->w + __vv*gp1->x*gp1->w)/tmp;
            pnt->y = (__uu*gp0->y*gp0->w + __vv*gp1->y*gp1->w)/tmp;
            pnt->z = (__uu*gp0->z*gp0->w + __vv*gp1->z*gp1->w)/tmp;
            pnt->w = (__uu*gp0->w + __vv*gp1->w)/tmp;
        }
    }
}

/*
 * RGREG2DU --- Obtain a partial derivative vector.
 *
 *  Parameters:
 *    rgp     (I)  Rational boundary Gregory patch
 *    i, j    (I)  Index
 *    u, v    (I)  Parameters
 *    pnt     (O)  Point with weight
 */
void
RGREG2DU(rgp, i, j, u, v, pnt)
    struct dbRational_Gregory_patch *rgp;
    int i, j;
    double u, v;
    struct dbHPoint *pnt;
{
    register struct dbHPoint *gp0, *gp1;
    double uu, vv, __uu, __vv;
    double tmp, tmp2;
    double u2, _u2;

    /*
     * Limit the degree of the rational boundary Gregory patch
     * to bicubic.
     */
    u2 = 2.0 * u;
    _u2 = 2.0 *(1.0 - u);
    gp0 = GP0(i,j);
    gp1 = GP1(i,j);
    if (i < 2) {
        vv = v * v;
        if (j < 2) {
            uu = u * u;
```

```
            tmp = (uu + vv) * (uu + vv);
            tmp2 = u2*vv/tmp;
            pnt->x = tmp2*(gp0->x*gp0->w - gp1->x*gp1->w);
            pnt->y = tmp2*(gp0->y*gp0->w - gp1->y*gp1->w);
            pnt->z = tmp2*(gp0->z*gp0->w - gp1->z*gp1->w);
            pnt->w = tmp2*(gp0->w - gp1->w);
        } else {
            __uu = (1.0 - u) * (1.0 - u);
            tmp = (__uu + vv) * (__uu + vv);
            tmp2 = _u2*vv/tmp;
            pnt->x = tmp2*(gp1->x*gp1->w - gp0->x*gp0->w);
            pnt->y = tmp2*(gp1->y*gp1->w - gp0->y*gp0->w);
            pnt->z = tmp2*(gp1->z*gp1->w - gp0->z*gp0->w);
            pnt->w = tmp2*(gp1->w - gp0->w);
        }
    } else {
        __vv = (1.0 - v) * (1.0 - v);
        if (j < 2) {
            uu = u * u;
            tmp = (uu + __vv) * (uu + __vv);
            tmp2 = u2*__vv/tmp;
            pnt->x = tmp2*(gp0->x*gp0->w - gp1->x*gp1->w);
            pnt->y = tmp2*(gp0->y*gp0->w - gp1->y*gp1->w);
            pnt->z = tmp2*(gp0->z*gp0->w - gp1->z*gp1->w);
            pnt->w = tmp2*(gp0->w - gp1->w);
        } else {
            __uu = (1.0 - u) * (1.0 - u);
            tmp = (__uu + __vv) * (__uu + __vv);
            tmp2 = _u2*__vv/tmp;
            pnt->x = tmp2*(gp1->x*gp1->w - gp0->x*gp0->w);
            pnt->y = tmp2*(gp1->y*gp1->w - gp0->y*gp0->w);
            pnt->z = tmp2*(gp1->z*gp1->w - gp0->z*gp0->w);
            pnt->w = tmp2*(gp1->w - gp0->w);
        }
    }
}
```

The partial derivative vector in the v direction can also be obtained by using the procedure above. Furthermore, by obtaining the outer product between two partial derivative vectors in the u and v directions at a parameter, the normal vector at the point on the surface can be obtained.

6. INTERSECTION CALCULATIONS

6.1 Introduction

Intersection calculations are basic calculations that make possible the 2-dimensional input, the ray tracing method, as well as the Boolean operations that are essential in 3-dimensional solid modelers. In Boolean operations, the intersection curves between surfaces have to be calculated, and in ray tracing, the intersection points between beams i.e., straight lines and surfaces must also be calculated. The intersection points between curves have also to be calculated in the 2-dimensional input.

Intersection calculations can be done by various methods. One is the algebraic method, another is the geometric method, or the Newton-Raphson method. At present, no method can be said to be the best as each method has its superior points. If a method is not practical for one purpose, it can be improved with some rearrangement. When implementing the program of an intersection calculation, the most suitable method should depend on the characteristics of the curves and surfaces. For example, to obtain the intersection point between a straight line and a circle, the algebraic method should be used. The geometric method or the Newton-Raphson method are better when calculating the intersection points between a free-form surface and a straight line. This chapter presents algorithms for various intersection calculations and explains the present problems with these algorithms.

6.2 Intersection Calculations Between Curves

The methods for calculating the intersection points between curves are: the algebraic method, the geometric method using division techniques, and the Newton-Raphson method.

6.2.1 Algebraic calculation method

In the popular method used to obtain the intersection points between curves such as arcs or straight lines, a parametric expression that represents the shapes is assigned to another implicitly-represented expression and then the resultant expression is solved.

Let us take an example of one method of calculating the intersection between a Bézier curve and a straight line on an XY-plane. The Bézier curve $\mathbf{P}(t) = (x(t), y(t))$ can be expressed by

$$x(t) = \sum_{i=0}^{n} B_i^n(t) x_i, \qquad y(t) = \sum_{i=0}^{n} B_i^n(t) y_i \tag{6.1}$$

where $B_i^n(t)$ is the Bernstein basis and $(x_i, y_i)(i = 0, \ldots, n)$ indicates the control points. The straight line can be expressed by

$$ax + by + c = 0 \tag{6.2}$$

where a, b, and c are real numbers (see Figure 6.1). By assigning Equation (6.1) to Equation (6.2),

$$a \sum_{i=0}^{n} B_i^n(t) x_i + b \sum_{i=0}^{n} B_i^n(t) y_i + c = 0$$

is obtained. This is an equation of degree n about t and can be solved algebraically up to $n = 4$.

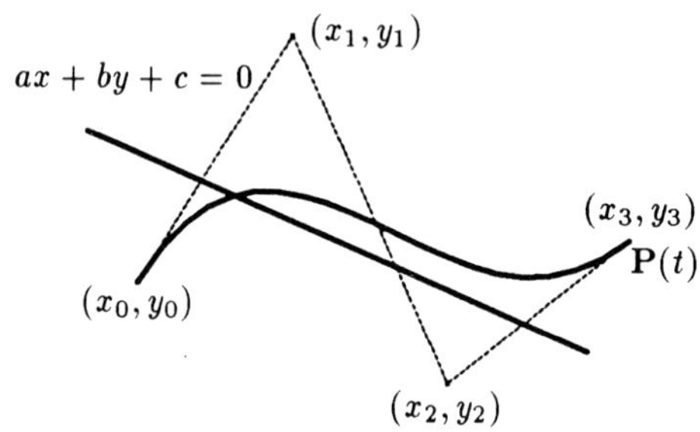

Fig. 6.1. Calculating the intersection between a Bézier curve and a straight line

The equation can easily be modified to obtain the variable t only, since the straight line is given by an equation of x and y. However, when calculating the intersection points between two Bézier curves $\mathbf{P}_1(t)$ and $\mathbf{P}_2(s)$, a single-variable equation cannot be easily obtained. $\mathbf{P}_1(t)$ and $\mathbf{P}_2(s)$ are given by

$$\mathbf{P}_1(t) = \sum_{i=0}^{n} B_i^n(t) \mathbf{P}_{1,i} \tag{6.3}$$

$$\mathbf{P}_2(s) = \sum_{i=0}^{n} B_i^n(s) \mathbf{P}_{2,i} \tag{6.4}$$

In order to solve Equations (6.3) and (6.4), one of the equations has to be changed into an implicit equation[57,126,128]. This method is called **implicitization**. We implicitize Equation (6.4) when $n = 2$ below. When $n = 2$, Equation (6.4) can be rewritten as

$$
\begin{aligned}
x &= a_2 s^2 + a_1 s + a_0 && (6.5)\\
y &= b_2 s^2 + b_1 s + b_0 && (6.6)
\end{aligned}
$$

Transposing the terms,

$$
\left.
\begin{aligned}
a_2 s^2 + a_1 s + (a_0 - x) &= 0\\
b_2 s^2 + b_1 s + (b_0 - y) &= 0
\end{aligned}
\right. \qquad (6.7)
$$

are obtained. Then implicitizing these equations by the Sylvester's method, they can be rewritten in the matrix representation as follows:

$$\mathbf{A}\mathbf{x} = 0$$

where

$$
\mathbf{A} = \begin{bmatrix} a_2 & a_1 & (a_0 - x) & 0 \\ 0 & a_2 & a_1 & (a_0 - x) \\ b_2 & b_1 & (b_0 - y) & 0 \\ 0 & b_2 & b_1 & (b_0 - y) \end{bmatrix}, \qquad \mathbf{x} = \begin{bmatrix} s^3 \\ s^2 \\ s^1 \\ 1 \end{bmatrix} \qquad (6.8)
$$

Because $\mathbf{x}$ has a solution other than $\mathbf{x} = [0,0,0,0]^T$, $|\mathbf{A}| = 0$ must be satisfied, and thus

$$
b_2^2 x^2 - 2a_2 b_2 xy + a_2^2 y^2 + (-2a_0 b_2^2 + a_1 b_1 b_2 - a_2 b_1^2 + 2a_2 b_0 b_2)x +
$$
$$
(-2b_0 a_2^2 + b_1 a_1 a_2 - b_2 a_1^2 + 2b_2 a_0 a_2)y + \begin{vmatrix} a_2 & a_1 & a_0 & 0 \\ 0 & a_2 & a_1 & a_0 \\ b_2 & b_1 & b_0 & 0 \\ 0 & b_2 & b_1 & b_0 \end{vmatrix} = 0
$$

is obtained. This is the form in which Equation (6.4) is expressed implicitly.

Sylvester's method implicitizes the Bézier curve of degree n with a $2n \times 2n$ matrix. Hence it is understandable that the implicitization of a Bézier curve of high degree is troublesome. The Bezout approach[22] makes the implicitization easier. To implicitize Equation (6.4) when $n = 2$, we must now define f and g as follows:

$$
\begin{aligned}
f &\stackrel{\text{def}}{=} a_2 s^2 + a_1 s + (a_0 - x)\\
g &\stackrel{\text{def}}{=} b_2 s^2 + b_1 s + (b_0 - y)
\end{aligned}
$$

The coefficients are simplified such that

$$
\begin{aligned}
f &= a_2' s^2 + a_1' s + a_0'\\
g &= b_2' s^2 + b_1' s + b_0'
\end{aligned}
$$

where

$$
\begin{aligned}
a_2' g - b_2' f &= (a_2' b_1' - a_1' b_2')s + (a_2' b_0' - a_0' b_2') &= 0\\
(a_2' s + a_1')g - (b_2' s + b_1')f &= (a_2' b_0' - a_0' b_2')s + (a_1' b_0' - a_0' b_1') &= 0
\end{aligned}
$$

These are rewritten into the matrix form as

$$\mathbf{A}\mathbf{x} = 0 \tag{6.9}$$

where

$$\mathbf{A} = \begin{bmatrix} (a'_2b'_1 - a'_1b'_2) & (a'_2b'_0 - a'_0b'_2) \\ (a'_2b'_0 - a'_0b'_2) & (a'_1b'_0 - a'_0b'_1) \end{bmatrix}, \quad \mathbf{x} = \begin{bmatrix} s \\ 1 \end{bmatrix} \tag{6.10}$$

Given the condition that $\mathbf{x}$ has a solution other than $\mathbf{x} = [0,0]^T$, then

$$|\mathbf{A}| = 0 \tag{6.11}$$

is obtained. This equation is expanded into

$$(a'_2b'_1 - a'_1b'_2)(a'_1b'_0 - a'_0b'_1) - (a'_2b'_0 - a'_0b'_2)(a'_2b'_0 - a'_0b'_2) = 0 \tag{6.12}$$

Now, Equation (6.4) is expressed as an implicit function. By assigning Equation (6.3) to Equation (6.12), an equation about t can be obtained. Here, t that satisfies $0 \leq t \leq 1$ is the parameter of an intersection point in Equation (6.3). If $s = t_0$, then according to Equation (6.4), a value of s that satisfies

$$\mathbf{P}_1(t_0) = \sum_{i=0}^{n} B_i^n(s)\mathbf{P}_{2,i}$$

is the parameter at an intersection point in Equation (6.4). This process is called **inversion**. When the degree of an equation is low, the value of s that satisfies the above equation can be obtained by finding the solution algebraically.

6.2.2 Geometric calculation method

Several methods have been suggested to calculate intersection geometrically. This subsection explains the **Bézier subdivision method**, the **interval subdivision method**[130], and the **Bézier clipping method** that use the characteristics of Bézier curves. It also explains the **geometric Newton-Raphson method** that does not depend on the characteristics or type of the curve.

Bézier subdivision method

The Bézier subdivision method obtains the intersection points between Bézier curves by using the convex hull property of Bézier curves and de Casteljau's algorithm for subdividing the curves (explained in Chapter 4). This method is accomplished through the following procedure:

1. From the control points, obtain the convex hulls of two Bézier curves (see Figure 6.2).

2. If the two convex hulls do not intersect, the two Bézier curves never intersect, thus terminate the processing (see Figure 6.3).

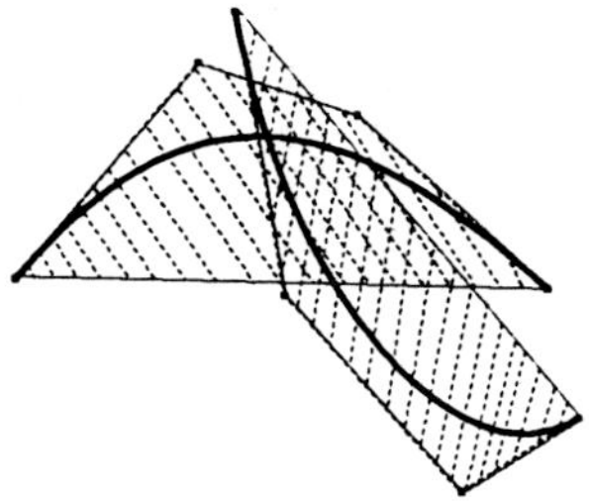

Fig. 6.2. Convex hulls of two Bézier curves

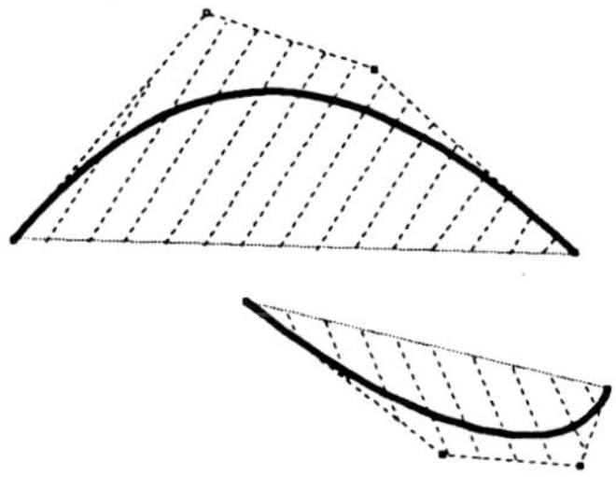

Fig. 6.3. No intersection between two convex hulls of Bézier curves

3. If the two Bézier curves can be regarded as almost straight lines, approximate the two curves to straight lines then obtain the intersection point. Otherwise, split each curve into two so that four Bézier curves are obtained, then repeat the steps from 1 to 3 (see Figure 6.4).

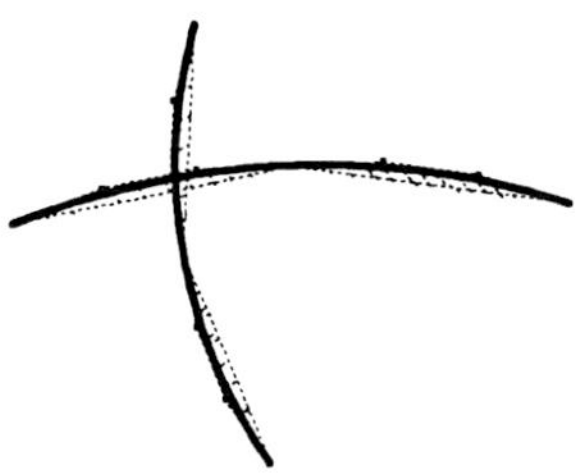

Fig. 6.4. Subdivision of two Bézier curves and convex hulls of the divided curves

In order to adjust the tolerance between the original curves and their approximated straight lines to lie within ϵ, Wang proposed the calculation method to obtain the number of divisions for the curves; this can be the upper limit in the case of the recursive calls[155]. If we define the control points of a Bézier curve of degree n on a plane to be $(x_i, y_i)(i = 0, \cdots, n)$ and

$$L_0 = \max_{0 \leq i \leq n-2} (|x_i - 2x_{i+1} + x_{i+2}|, |y_i - 2y_{i+1} + y_{i+2}|)$$

is set, then the number of divisions r_0 is set to

$$r_0 = \log_4 \frac{\sqrt{2}n(n-1)L_0}{8\epsilon}$$

This method can be easily implemented and a stable solution can easily be obtained.

Interval subdivision method

The interval subdivision method divides a Bézier curve into intervals, each of whose end points have horizontal or vertical tangents. Figure 6.5(a) illustrates a Bézier curve before subdivision and (b) illustrates the curve after subdivision. All intervals form rectangular convex hulls. Figure 6.6(a) illustrates a convex hull of a Bézier curve before subdivision and (b) illustrates the convex hulls after subdivision. After dividing a Bézier curve into several intervals, the same procedure as that outlined in the Bézier subdivision method can be applied. Figure 6.7(a) illustrates two Bézier curves intersecting and (b) illustrates intervals created from the curves and their convex hulls. The interval subdivision method is faster than the Bézier subdivision method but the implementation is more complicated. A stable solution can be readily obtained.

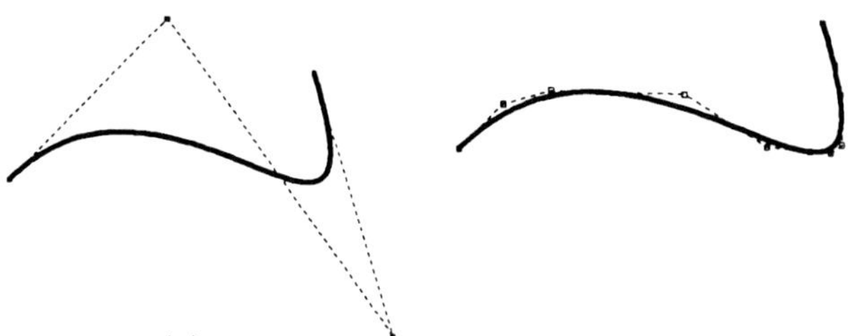

(a) (b)

Fig. 6.5. Division of a Bézier curve for interval subdivision method

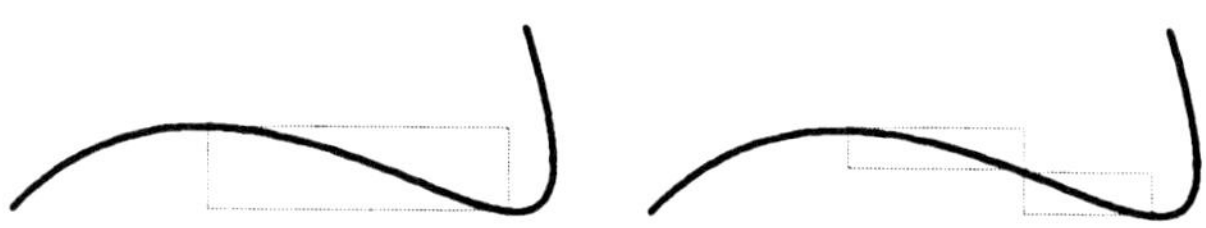

(a) (b)

Fig. 6.6. Convex hulls for interval subdivision method

The **hodograph** is helpful to divide curves at points where horizontal and vertical tangents cross. For a Bézier curve of degree n, $\mathbf{P}(t) = \sum_{i=0}^{n} B_i^n(t)\mathbf{P}_i$, the hodograph is defined by

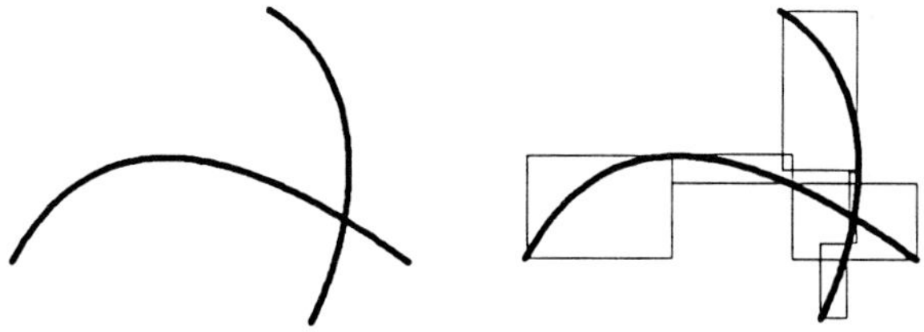

(a) (b)

Fig. 6.7. Intervals and their convex hulls in the interval subdivision method

$$\mathbf{H}(t) = \frac{\partial \mathbf{P}(t)}{\partial t} = n \sum_{i=0}^{n-1} B_i^{n-1}(t)(\mathbf{P}_{i+1} - \mathbf{P}_i)$$

This has interesting characteristics, namely if we define the parameter of an intersection point between $\mathbf{H}(t)$ and the x-axis to be t_{x0}, then $\mathbf{P}(t)$, where $t = t_{x0}$, is the extremum in the x-axis direction. Likewise, with the parameter t_{y0} in the case where $\mathbf{H}(t)$ and the y-axis intersect, $\mathbf{P}(t)$ is the extremum in the y-axis direction. The hodograph of the rational Bézier curve can also be obtained easily[12,131].

Bézier clipping method

At SIGGRAPH in 1990, Nishita et al. proposed a new algorithm to calculate the intersection between a rational Bézier patch and a straight line[101]. By modifying this algorithm, intersection points between a rational Bézier curve and a straight line can be obtained[100,99]. This application can be regarded as an improved Bézier subdivision method. In the Bézier subdivision method, a Bézier curve is divided into two at the 0.5 parameter so that the convex hull of the curve is made smaller. On the other hand, in the Bézier clipping method, the curves are further divided depending on where they intersect.

When a Bézier curve and a straight line lie on the XY-plane, the coordinate system can easily be translated so as to meet $y = y'$ with the straight line. If, as a result of the translation, a point on the Bézier curve can be found such that its tangent is parallel to the x-axis, then the curve is divided, using the hodograph, at the point. The Bézier curve $\mathbf{P}(t)$ obtained by this operation is given by

$$\mathbf{P}(t) = \sum_{i=0}^{n} B_i^n(t)\mathbf{P}_i \tag{6.13}$$

where $\mathbf{P}_i = (x_i, y_i)(i = 0, \cdots, n)$ indicates the control points of the Bézier curve. The intersection point between $y = y'$ and Equation (6.13) can be calculated using

$$\sum_{i=0}^{n} B_i^n(t) y_i = y' \tag{6.14}$$

Equation (6.14) is modified to give

$$\sum_{i=0}^{n} B_i^n(t) d_i = 0 \tag{6.15}$$

where $d_i = y_i - y'$. Here, Equation (6.15) is set to $f(t)$, which is called the **Bézier function**[53]. A Bézier curve $\mathbf{d}(t)$ formed by the control points $(i/n, d_i)$ is given by

$$\mathbf{d}(t) = \sum_{i=0}^{n} B_i^n(t) \begin{bmatrix} \frac{i}{n} \\ d_i \end{bmatrix} \tag{6.16}$$

where $\mathbf{d}(t)$ is equal to the distance from $y = y'$ to the Bézier curve. This $\mathbf{d}(t)$ is called the **distance curve**.

The method to obtain t which satisfies $\mathbf{d}(t) = 0$ is known as the **Bézier clipping method**. Referring to Figure 6.8(a) which shows the intersection point between a Bézier curve and $y = y'$, the steps are as follows:

1. Obtain the convex hull of $\mathbf{d}(t)$.
2. Calculate t_{min} and t_{max}, the minimum and maximum values of t at the intersection point between the convex hull and t. (Figure 6.8(b)). If $|t_{\text{max}} - t_{\text{min}}|$ is within the tolerance, terminate the processing.
3. Divide $\mathbf{d}(t)$ by the interval $[t_{\text{min}}, t_{\text{max}}]$ and set the result obtained to $\mathbf{d}(t)$, then repeat the steps from 1.

This method, although limited to calculations of the intersection between a Bézier curve and a straight line, is as stable as the Bézier subdivision method and its execution can be carried out much faster.

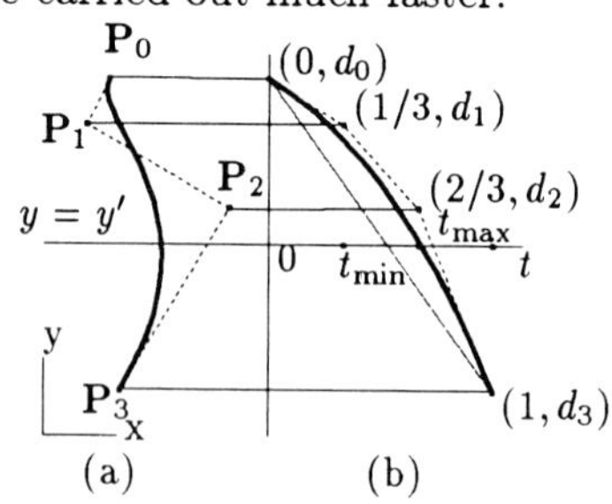

Fig. 6.8. Bézier clipping method

Geometric Newton-Raphson method

The precision of a solution obtained by the Bézier subdivision method can be improved by using the derivation of a curve. We now define two Bézier curves as $\mathbf{P}_1(t)$ and $\mathbf{P}_2(s)$ and the rough parameters of the intersection point of the two curves as t_0 and s_0. Then, the intersection point of the tangents at t_0 and s_0 is closer to the precise intersection point than $\mathbf{P}_1(t_0)$ and $\mathbf{P}_2(s_0)$ (Figure 6.9). This algorithm can be used to obtain an intersection point between two curves:

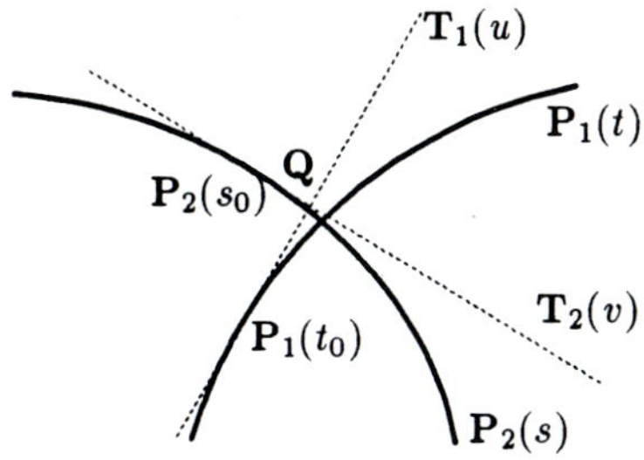

Fig. 6.9. Intersection calculation between curves by Newton-Raphson method

1. Obtain $\mathbf{P}_1(t_0)$, $\mathbf{P}_2(s_0)$, $\frac{\partial \mathbf{P}_1(t_0)}{\partial t}$, and $\frac{\partial \mathbf{P}_2(s_0)}{\partial s}$.
2. If $\mathbf{P}_1(t_0)$ and $\mathbf{P}_2(s_0)$ have the same coordinates, the parameters t_0 and s_0 are the parameters of the intersection point.
3. Obtain the intersection point $\mathbf{Q}$ between two tangents at points $\mathbf{P}_1(t_0)$ and $\mathbf{P}_2(s_0)$: $\mathbf{T}_1(u) = \mathbf{P}_1(t_0) + u\frac{\partial \mathbf{P}_1(t_0)}{\partial t}$ and $\mathbf{T}_2(v) = \mathbf{P}_2(s_0) + v\frac{\partial \mathbf{P}_2(s_0)}{\partial s}$.
4. Obtain $\mathbf{v}_1 = \mathbf{Q} - \mathbf{P}_1(t_0)$ and $\mathbf{v}_2 = \mathbf{Q} - \mathbf{P}_2(s_0)$ and solve
$$\mathbf{v}_1 = \delta t \frac{\partial \mathbf{P}_1(t_0)}{\partial t}, \qquad \mathbf{v}_2 = \delta s \frac{\partial \mathbf{P}_2(s_0)}{\partial s}$$
5. Update the values of t_0 and s_0 as follows:
$$t_0 = t_0 + \delta t, \qquad s_0 = s_0 + \delta s$$
then repeat the steps from 1.

If the coordinates and the first-order derivative can be obtained about two curves, the geometric Newton-Raphson method can be used to obtain the intersection point between them regardless of the type of curve. If the tangents of the curves lie almost in parallel, however, the convergence will be slow and unstable.

The processes here are almost similar to the Newton-Raphson method, and so we call this method **geometric Newton-Raphson method** in the following explanations.

6.3 Intersection Calculations Between a Curve and a Surface

This section explains how to calculate intersection points between a curve and a surface. When a curve is straight line or arc type and a surface is a natural quadric, the algebraic calculation method is useful. However, the calculation becomes complicated when determining the intersection points between a free-form curve and a free-form surface. In this case, the Newton-Raphson method is suitable. This section outlines the intersection calculations based on both the algebraic and geometric Newton-Raphson methods.

6.3.1 Algebraic calculation method

Just as the intersection points between curves are obtained by the algebraic calculation method, the intersection points between a free-form curve and a free-form surface can also be obtained algebraically. In this calculation, however, accurate solutions cannot always be obtained when the degree of algebraic expressions is high. Sederberg et al. proposed a method to obtain the intersection points between a straight line and the patch algebraically[127] by using a quadratic triangular surface patch called the **Steiner patch**. This method however can hardly be applied to general cases. Among the intersection calculations, the algebraic method is the most effective for obtaining the intersection points between a straight line and a natural quadric surface. For example, the intersection points between a straight line $\mathbf{P}(t) = \mathbf{P}_0 + \mathbf{v}t$ and a sphere $x^2 + y^2 + z^2 = r^2$ can be obtained as follows (see Figure 6.10):

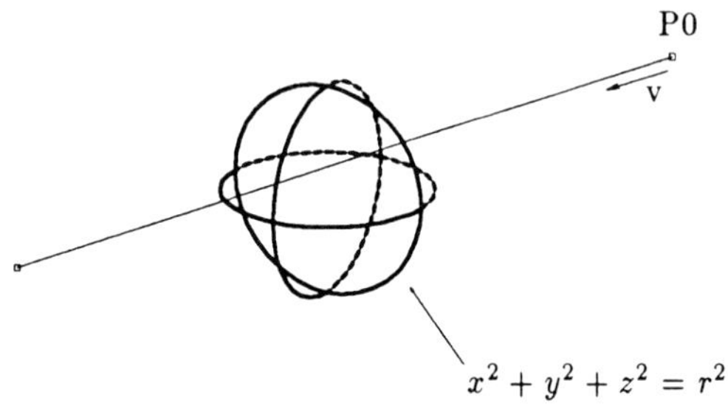

Fig. 6.10. Intersection calculation between a straight line and a sphere

Rewrite the equation of the straight line as

$$\begin{bmatrix} x(t) \\ y(t) \\ z(t) \end{bmatrix} = \begin{bmatrix} \mathbf{P}_{0x} \\ \mathbf{P}_{0y} \\ \mathbf{P}_{0z} \end{bmatrix} + t \begin{bmatrix} \mathbf{v}_x \\ \mathbf{v}_y \\ \mathbf{v}_z \end{bmatrix}$$

and assign it to the equation of the sphere, such that

$$(\mathbf{P}_{0x} + t\mathbf{v}_x)^2 + (\mathbf{P}_{0y} + t\mathbf{v}_y)^2 + (\mathbf{P}_{0z} + t\mathbf{v}_z)^2 = r^2$$

This equation is a quadratic exression about t and can be easily solved. The intersection points between a circle and a natural quadric surface can be obtained in the same way by solving a quartic expression. To obtain the intersection point between a cubic Bézier surface and a natural quadric surface, a 6th-degree expression has to be solved.

6.3.2 Geometric calculation method

The geometric calculation method is frequently used to obtain the intersection points between a free-form curve and a free-form surface. This subsection explains how to calculate the intersection points by applying the geometric Newton-Raphson method.

We now define a C^2 continuous parametric curve as $\mathbf{P}(t)$ and a C^2 continuous parametric patch as $\mathbf{S}(u,v)$. The intersection points between the curve and the patch can be obtained from the following procedure:

1. Perform a rough check for the intersection. If no evidence of intersection is found, terminate the operation.
2. Approximate the curve with several line segments and approximate the surface with several triangular surface patches.
3. Obtain the intersection points between the elongated line segments and infinite planes on which the patches are lying. If the parameters of the intersection points obtained are not far from the parameter space of the patch, then they can be considered an approximation of the intersection points.
4. Adjust each of the approximation points so as to obtain the precise intersection points.

Note that in step 3, the approximation points cannot always be obtained. Take the example, when a straight line and a surface intersect as shown in Figure 6.11. If the surface is approximated with triangular patches as shown in Figure 6.12, no intersection points can be obtained between the line and the patch B but they can be obtained between the line and the patches A and C.

We define the parameters of the approximation points on a curve $\mathbf{P}(t)$ to be t_0 and those on a surface $\mathbf{S}(u,v)$ to be u_0 and v_0 (see Figure 6.13). To adjust approximation points, perform the following steps:

1. Calculate $\mathbf{P}(t_0)$ and $\frac{\partial \mathbf{P}(t_0)}{\partial t}$, so that a straight line l is obtained which passes through the point $\mathbf{P}(t_0)$ and has the direction vector $\frac{\partial \mathbf{P}(t_0)}{\partial t}$.
2. Calculate $\mathbf{S}(u_0,v_0)$, $\frac{\partial \mathbf{S}(u_0,v_0)}{\partial u}$, and $\frac{\partial \mathbf{S}(u_0,v_0)}{\partial v}$, so that a plane P is obtained to pass through $\mathbf{S}(u_0,v_0)$ and has the normal vector $\frac{\partial \mathbf{S}(u_0,v_0)}{\partial u} \times \frac{\partial \mathbf{S}(u_0,v_0)}{\partial v}$.
3. Obtain the intersection point $\mathbf{Q}$ between the straight line l and the plane P. Then obtain the vectors $\mathbf{v}_1$ from $\mathbf{P}(t_0)$ to $\mathbf{Q}$ and $\mathbf{v}_2$ from $\mathbf{S}(u_0,v_0)$ to $\mathbf{Q}$.
4. Obtain δt by solving $\mathbf{v}_1 = \delta t \frac{\partial \mathbf{P}(t_0)}{\partial t}$.
5. Obtain δu and δv by solving $\mathbf{v}_2 = \delta u \frac{\partial \mathbf{S}(u_0,v_0)}{\partial u} + \delta v \frac{\partial \mathbf{S}(u_0,v_0)}{\partial v}$.

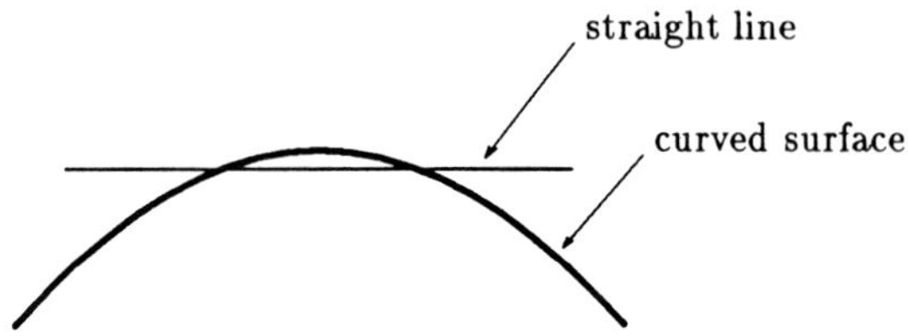

Fig. 6.11. Intersection between a straight line and a surface (cross-section)

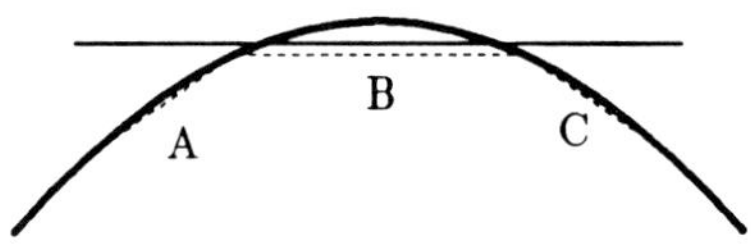

Fig. 6.12. Intersection between a straight line and approximated triangles A, B, and C (cross-section)

6. If δt, δu, and δv are regarded to be 0, then terminate the processing.
7. Define $t_0 = t_0 + \delta t$, $u_0 = u_0 + \delta u$, and $v_0 = v_0 + \delta v$, and go back to the step 1.

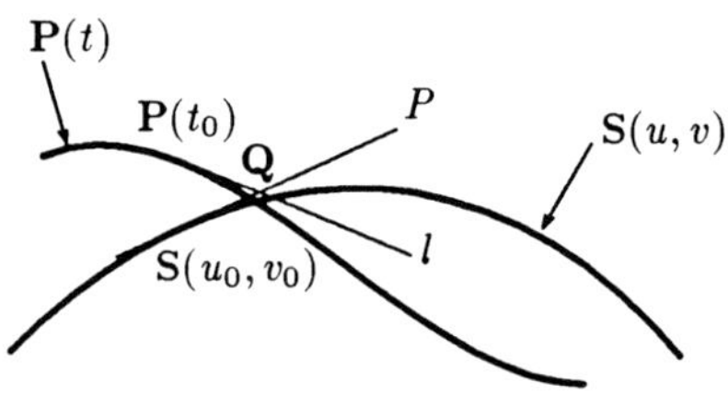

Fig. 6.13. Intersection calculation between a curve and a surface by geometric Newton-Raphson method (cross-section)

This method can obtain intersection points by using coordinates and derivative vectors to which parameters are applied, and does not depend on the representations of curves and surfaces.

6.4 Intersection Calculations Between Surfaces

Among the operations to create 3-dimensional solid models, the most difficult to execute completely are the Boolean operations. The difficulty with the Boolean operations is caused by two factors: one, is that the topology of a solid must be precisely recreated but this can become complicated, and two is that the correct intersection curves between surfaces cannot be completely obtained[111]. This section explains how to calculate the intersection curves between surfaces.

The intersection calculations between surfaces can be classified into two categories: one is the calculation between quadratic surfaces, and the other involves calculations between free-form surfaces, between a quadratic surface and a free-form surface, and between a plane and a free-form surface.

6.4.1 Intersection calculations between quadratic surfaces

The method proposed by Levin[87,86] to algebraically obtain the intersection curves between quadratic surfaces was actually applied in solid modelers[124]. After that, Miller proposed another method to obtain the shape of the intersection curves using the locational relationship of two quadratic surfaces[97,137].

Algebraic calculation method

Let us first consider the method for algebraically obtaining the intersection curves between quadratic surfaces. The general form of a quadratic surface is given by:

$$q(x,y,z) = Ax^2 + By^2 + Cz^2 + 2Dxy + 2Eyz + 2Fxz + 2Gx + 2Hy + 2Jz + K$$

where $A, B, \ldots,$ and K are constants. This equation can be written in the matrix form as

$$q(x,y,z) = \begin{bmatrix} x & y & z & 1 \end{bmatrix} \begin{bmatrix} A & D & F & G \\ D & B & E & H \\ F & E & C & J \\ G & H & J & K \end{bmatrix} \begin{bmatrix} x \\ y \\ z \\ 1 \end{bmatrix} \equiv \mathbf{x}Q\mathbf{x}^T$$

where Q is a 4×4 matrix. Here, quadratic surfaces defined by the coefficient matrices Q_1 and Q_2 are given by

$$\mathbf{x}Q_1\mathbf{x}^T = 0, \qquad \mathbf{x}Q_2\mathbf{x}^T = 0$$

Suppose α is an arbitrary real number,

$$\mathbf{x}(Q_1 - \alpha Q_2)\mathbf{x}^T = 0$$

indicates the **family** of quadratic surfaces called a **pencil**. Each pencil is a quadratic surface that passes through the intersection curve between Q_1 and Q_2 (if they intersect).

Levin proved that a pencil of two quadratic surfaces has at least one **ruled quadratic surface**. The ruled quadratic surface is one that contains a point at which the intersection between the surface and a plane passing the point forms a straight line. A ruled quadratic surface can be parameterized such that its intersection line becomes:

$$a(t)s^2 + b(t)s + c(t) = 0 \tag{6.17}$$

where t and s are the parameters and s is the distance from the reference point on the quadratic surface to the point on the ruled quadratic surface along the generating line.

When a quadratic surface forms a cylinder $x^2 + y^2 = r^2$, the functions $a(t)$, $b(t)$, and $c(t)$ are given by

$$\begin{aligned} a(t) &= C \\ b(t) &= 2Er\sin t + 2Fr\cos t + 2J \\ c(t) &= Ar^2\cos^2 t + Br^2\sin^2 t + 2Dr^2\sin t\cos t \\ &\quad +2Gr\cos t + 2Hr\sin t + K \end{aligned}$$

When it forms a cone $x^2 + y^2 - m^2z^2 = 0$,

$$\begin{aligned} a(t) &= Am^2\cos^2 t + Bm^2\sin^2 t + C + 2Dm^2\sin t\cos t \\ &\quad +2Em\sin t + 2Fm\cos t \\ b(t) &= 2Gm\cos t + 2Hm\sin t + 2J \\ c(t) &= K \end{aligned}$$

The intersection curves between quadratic surfaces can be obtained as follows:

1. Obtain $a(t)$, $b(t)$, and $c(t)$.
2. Obtain t: one when s is ∞, one at an end point of an intersection curve, and one at the point where the intersection curve bends, i.e., t that satisfies the discriminant $b(t)^2 - 4a(t)c(t) = 0$. These points are called **critical points**.
3. Obtain an equation by varying t of Equation (6.17) between two critical points and solve the equation about s so that points on the intersection curve can be found.
4. Create a curve by interpolating the obtained points.

The algebraic calculation method contains problems in that it becomes unstable if the coefficients of Equation (6.17) are changed slightly. This small change causes the position of the point that satisfies this equation (a point on the intersection curve) to change greatly, and each coefficient of the discriminant $b(t)^2 - 4a(t)c(t)$ then contains errors. Therefore, the criterion used to determine that the solution is 0 might be ambiguous. These problems occur because quadratic surfaces have to be defined by quadratic expressions and are solved using these expressions. The geometric method is proposed to compensate for these problems.

Geometric calculation method

The geometric calculation method attempts to define the shape of intersection curves from the locational relationship between two natural quadric surfaces. The shape of a sphere, cylinder, and cone can be defined using the following values and vectors (see Figure 6.14):

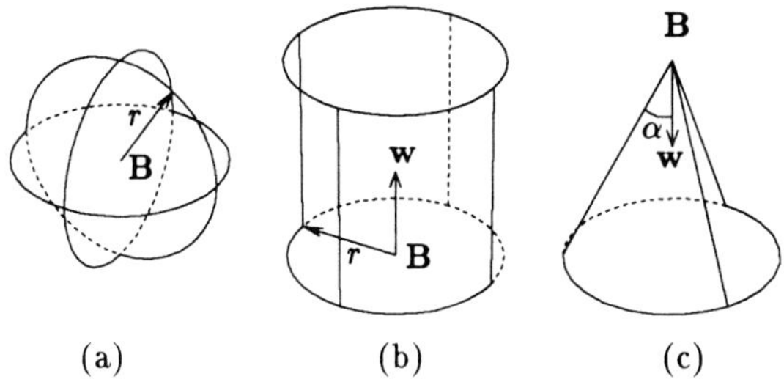

Fig. 6.14. Natural quadric surfaces (sphere, cylinder, cone)

Sphere: center **B**, radius r
Cylinder: point on the axis **B**, unit axis vector **w**, radius r A cylinder
Cone: vertex **B**, unit axis vector **w**, vertical angle α
and cone also have two vectors **u** and **v**, which are orthogonal both to **w** and to each other.

Here, the parametric representation of the cylinder is given by

$$\mathbf{P}(s,t) = \gamma(t) + s\mathbf{w} \tag{6.18}$$

where

$$\gamma(t) = \mathbf{B} + r(\mathbf{u}\cos t + \mathbf{v}\sin t)$$

With an arbitrary t, $\gamma(t)$ is a point on the circle of center **B**, normal vector **w**, and radius r.

The parametric representation of a cone is given by

$$\mathbf{P}(s,t) = \mathbf{B} + s(\delta(t) + \mathbf{w}) \tag{6.19}$$

where

$$\delta(t) = \tan\alpha(\mathbf{u}\cos t + \mathbf{v}\sin t)$$

With an arbitrary t, $\delta(t)$ is a vector of length $\tan\alpha$, orthogonal to **w**.

When **P** is a position variable, a sphere, cylinder, and cone are implicitized as

$$\text{Sphere:} \quad (\mathbf{P}-\mathbf{B})\cdot(\mathbf{P}-\mathbf{B}) - r^2 = 0 \tag{6.20}$$

$$\text{Cylinder:} \quad (\mathbf{P}-\mathbf{B})\cdot(\mathbf{P}-\mathbf{B}) - ((\mathbf{P}-\mathbf{B})\cdot\mathbf{w})^2 - r^2 = 0 \tag{6.21}$$

$$\text{Cone:} \quad ((\mathbf{P}-\mathbf{B})\cdot\mathbf{w})^2 - \cos^2\alpha(\mathbf{P}-\mathbf{B})\cdot(\mathbf{P}-\mathbf{B}) = 0 \tag{6.22}$$

By assigning Equations (6.18) or (6.19) to Equation (6.20), (6.21), or (6.22),

$$\mathbf{a}(t)s^2 + \mathbf{b}(t)s + \mathbf{c}(t) = 0 \tag{6.23}$$

is obtained. Equation (6.23) expresses the intersection curves between parametric natural quadric surfaces. The intersection curves between spheres can easily be obtained, hence no implicitization is necessary.

Figure 6.15 illustrates two cylinders intersecting. The geometric method classifies the shape of the intersection curves by the positional relationship between the natural quadric surfaces and then obtains the critical points. In the example of Figure 6.15, the critical points can be defined when $t = t_0$, $t = t_c$, and $t = t_1$. s in this case can be obtained by assigning t to Equation (6.23) such that it defines the critical point. Then, between critical points, by assigning proper t to Equation (6.23), s for the given t is obtained. For example, in Figure 6.15, $s = s_0$ and $s = s_1$ are obtained by assigning $t = t'$ in Equation (6.23). Between $t = t_0$ and $t = t'$, two intersection curves can be found in the s direction. Hence by approximating the obtained points to a curve, the intersection curve can be obtained.

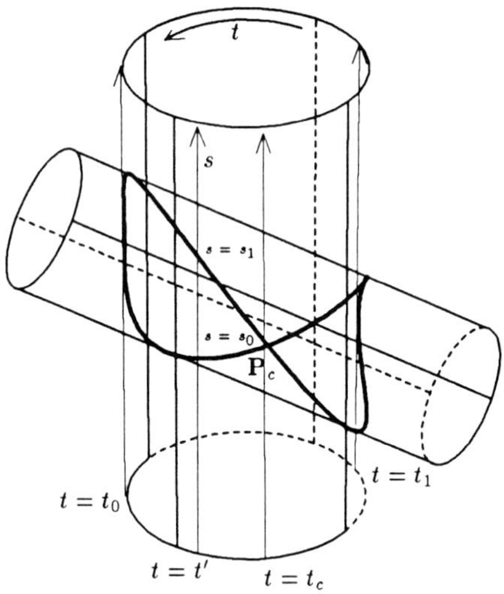

Fig. 6.15. Two cylinders intersecting

Next, let us consider the classification of the intersection curves between a sphere and a cylinder and between a sphere and a cone. The classifications of the intersection curves between cylinders, between a cylinder and a cone, and between cones are omitted because of a limitation of pages. For the detailed explanation, please refer to the paper by Miller[97].

- **sphere/cylinder**
  The radius of a sphere is defined as r_s and the radius of the base of a cylinder is defined as r_c. The shortest distance between the center of the sphere and the axis of the cylinder is defined as d. Here, the intersection curves can be classified as follows:

$d = 0$ &	
$r_s < r_c$	No intersection
$r_s = r_c$	A circle
$r_s > r_c$	Two circles (Figure 6.16(a))
$d \neq 0$ &	
$d > r_s + r_c$	No intersection
$d = r_s + r_c$	Tangent on a point
$d < r_s + r_c$ &	
$r_s < r_c$ &	
$d < r_c - r_s$	No intersection
$d = r_c - r_s$	A sphere inscribing in a cylinder
$d > r_c - r_s$	A closed curve (Figure 6.16(b))
$r_s \geq r_c$ &	
$d < r_s - r_c$	Two closed curves (Figure 6.16(c))
$d = r_s - r_c$	An 8-figure closed curve (Figure 6.16(d))
$d > r_s - r_c$ &	
$d^2 \leq {r_s}^2 - {r_c}^2$	A closed curve (Figure 6.16(e))
$d^2 > {r_s}^2 - {r_c}^2$	A closed curve (Figure 6.16(f))

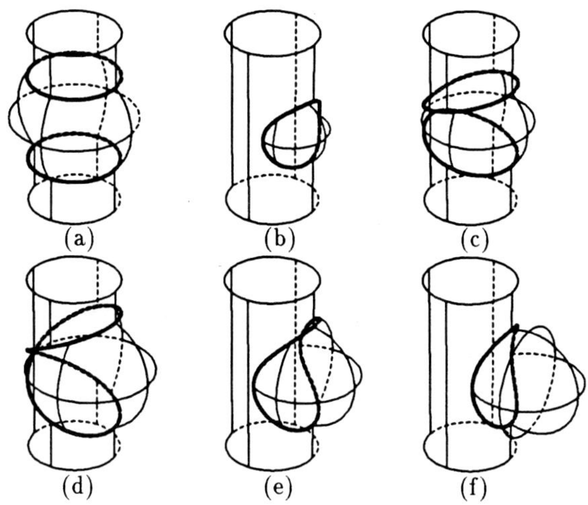

Fig. 6.16. Intersection curves between a sphere and a cylinder

- **sphere/cone**
  The radius of a sphere is defined as r, α is the tangent of the vertical angle of a cone, d is the length of the perpendicular between the center of the sphere and the axis of the cone, and h, the length of the orthogonal projection of the vector from the vertex of the cone to the center of the sphere onto the axis of the cone. Offset$=r\sqrt{1+\alpha^2}/\alpha$ is the value used

to check whether the sphere lies inside or outside of the cone. Here, the intersection curves can be classified as follows:

$\sqrt{h^2+d^2} < r$	A closed curve (Figure 6.17(a))
$\sqrt{h^2+d^2} = r$	A closed curve and a vertex
$\sqrt{h^2+d^2} > r$ &	Vertex lying outside of sphere
$\alpha d \leq -h$	Sphere lies under cone. If equal, vertex lies on sphere.
$\alpha d > -h$ &	Checks whether sphere lies inside or outside of cone.
$d \geq \alpha(h + \text{offset})$	Sphere lies outside of cone. If equal, sphere is tangent to cone at one point.
$d \leq \alpha(h - \text{offset})$	Sphere lies inside of cone If equal, sphere is tangent to cone at one point.
$d < \alpha(\text{offset} - h)$	Two closed curves (Figure 6.17(b))
Others	A closed curve (Figure 6.17(c)(d))

Figure 6.17 illustrates how the intersection curves are created.

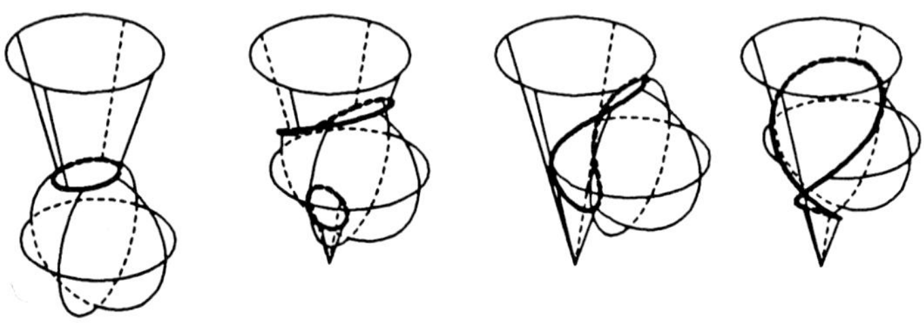

(a) (b) (c) (d)

Fig. 6.17. Intersection curves between a sphere and a cone

Despite the complicated classification, the geometric calculation method is the major method used for intersection calculations because of its stability and its efficiency.

6.4.2 Intersection calculations between free-form surfaces

It is very difficult to obtain the intersection curves between free-form surfaces. For example, the intersection curves between two bicubic Bézier patches are of the 324th degree. This can prove to be difficult when obtaining the intersection curves algebraically. Because of this difficulty, the current methods give an approximation for the intersection curves. At present, two major methods are used — the **recursive subdivision method**[50,69,80,82,105] and the **marching method**[4,5,28].

The recursive subdivision method divides free-form surfaces recursively and calculates the intersection between the **subpatches** generated. The marching method obtains one point on an intersection curve and then traces the intersection curve from this point.

6.4.3 Recursive subdivision method

The recursive subdivision method is an expansion of the Bézier subdivision method described in Section 6.2. This method can be applied to any patch once it can be divided into subpatches and has the convex hull property. The intersection curves between patches $\mathbf{S}_1$ and $\mathbf{S}_2$ are obtained by the following procedure:

1. Obtain the convex hulls of the patches $\mathbf{S}_1$ and $\mathbf{S}_2$ and perform a rough check on the convex hulls obtained to see whether they intersect. If they do not intersect, terminate the processing.
2. Subdivide each of the patches $\mathbf{S}_1$ and $\mathbf{S}_2$ into four subpatches. Obtain the convex hulls for each subpatch, and check for intersection of the convex hulls generated. Repeat this operation on subpatches whose convex hulls intersect. Figure 6.18 illustrates intersecting subpatches which have been recursively divided.

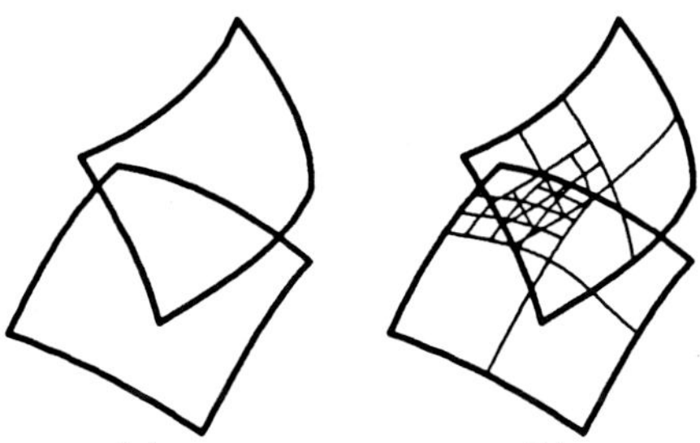

(a) (b)

Fig. 6.18. Recursive subdivision of surfaces

Store the data on the subpatches generated in the **quadtree structure** (Figure 6.19).

3. Each subpatch can be represented by two triangular patches. The intersection curve of the original patch can be roughly obtained from the intersection lines of the triangular patches. Figure 6.20(a) illustrates subpatches represented by triangular patches and (b) shows their rough intersection curves. We call these intersection curves **line segments**.
4. Connect the line segments so as to create a list i.e. line segment list that consists of the several intersection lines.

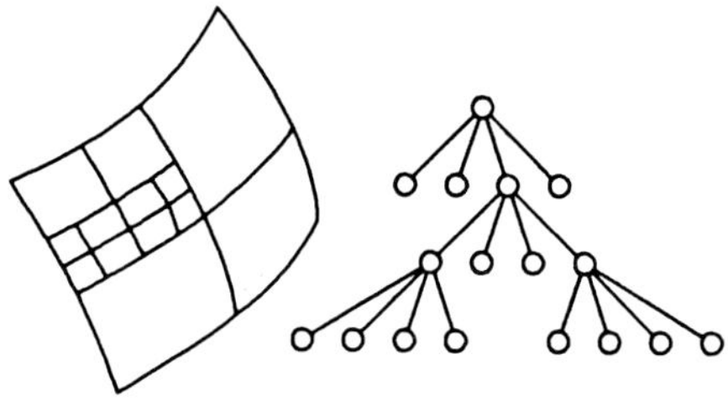

Fig. 6.19. A divided patch and its data in a quadtree

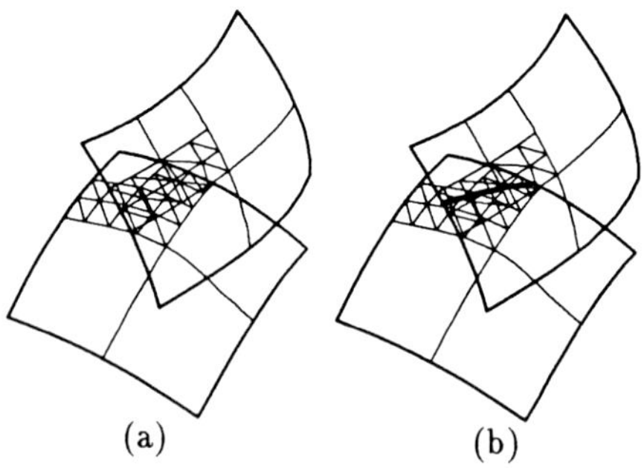

Fig. 6.20. Intersection curves between patches approximated by triangular patches

If a patch is parametrically divided, the subpatches (as shown in Figure 6.20) can be regarded as rectangular areas as illustrated in Figure 6.21. When this area contains a line segment AB, as shown in Figure 6.22, check the upper subpatch from A and the right-side patch from B to find line segments which connect to the AB.

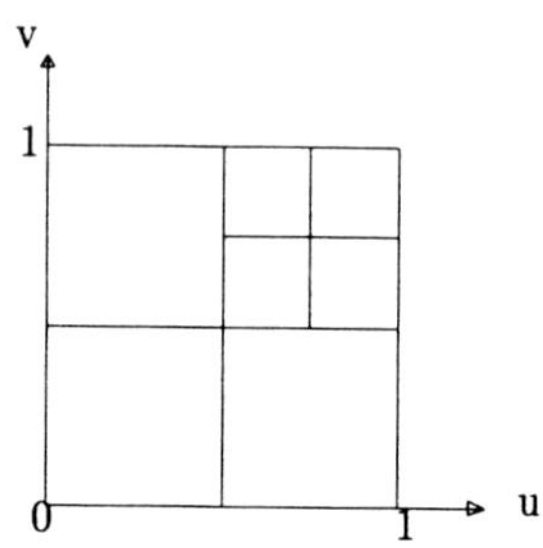

Fig. 6.21. A patch divided in the parametric space

5. Because each line segment has error, refine the end points of a line segment using the Newton-Raphson method. Then, select several points on the line segment and refine the points by the Newton-Raphson method (see Figure 6.23) so that the precise intersection curve can be obtained.

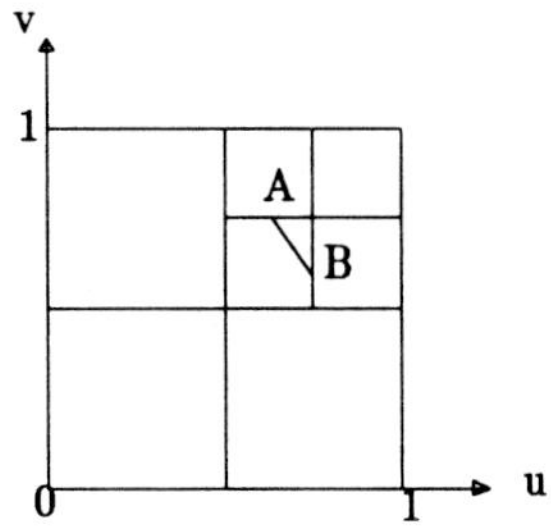

Fig. 6.22. A line segment in the parametric space

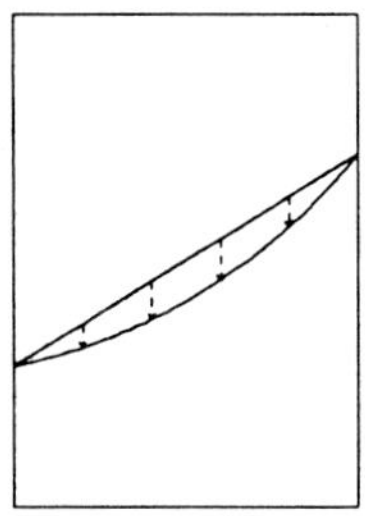

Fig. 6.23. Calculation of precise points on an intersection curve

Dividing a Bézier patch, rational Bézier patch, or NURBS is made easier by using the de Casteljau's algorithm, described in Chapter 4. Gregory patches and/or rational boundary Gregory patches cannot be divided, but they can be converted into rational Bézier patches[142]. The recursive subdivision method applies to patches that can be divided into several subpatches and have the convex hull property.

6.4.4 Marching method

The recursive subdivision method divides a surface into subpatches, so that if the surface shape is simple the intersection curves can be obtained. However, in the case of a trimmed surface or when intersection curves intersect themselves, they cannot be accurately obtained. The marching method[93] with its fast processing can be used in such situations. The marching method obtains intersection curves by tracing the intersection from a point. This method has one drawback in that it cannot obtain completely an intersection curve unless the starting point for tracing is found. However, this problem is being resolved through several studies[29,93,129].

The starting point for tracing can be found by calculating the intersection between a surface and a curve or by calculating one of the internal intersection points. The point obtained is registered in a table (called the **end point table**) and the tracing starts from this point. The point registered in the end point table is the possible starting point and at the same time the possible termination point. By using the end point table, intersection curves between trimmed surfaces can be easily obtained.

In Figure 6.24 the intersection curves between the parametric surfaces $\mathbf{S}^1$ and $\mathbf{S}^2$, are obtained through the following steps:

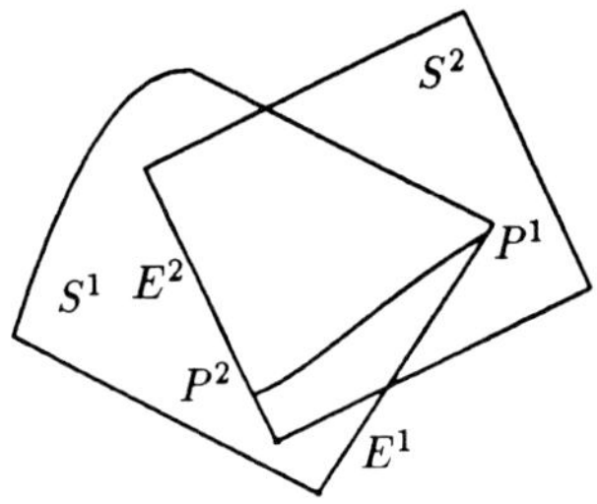

Fig. 6.24. Calculating intersection between surfaces using the marching method

1. Perform the rough check for intersections. If no intersection can be found, terminate the processing.

2. Obtain the intersection point $\mathbf{P}^1$ between a boundary curve E^1 of a surface $\mathbf{S}^1$ and a surface $\mathbf{S}^2$ and register the point obtained in the end point table.

   (a) Obtain the intersection point $\mathbf{P}^1$ between E^1 and $\mathbf{S}^2$ and calculate the parameter of $\mathbf{P}^1$ on the surface $\mathbf{S}^2$.

   (b) Calculate the parameter of the intersection point $\mathbf{P}^1$ on the surface $\mathbf{S}^1$.

   (c) An edge created as a result of surface-surface intersection may not always lie exactly on its original surface[25]. For the purposes of this book, we shall call such an edge a **trimming edge**. If E^1 is a trimming edge, obtain the intersection point between three surfaces, i.e., the adjacent two surfaces lying on E^1 and the surface $\mathbf{S}^2$, and regard this point as being the exact intersection point between E^1 and $\mathbf{S}^2$. Otherwise, the intersection point obtained in the step 2a is the intersection point.

   (d) Register the parameters of the intersection point on each of the surfaces in the end point table.

3. Obtain the intersection point $\mathbf{P}^2$ between a boundary curve E^2 of the surface $\mathbf{S}^2$ and the surface $\mathbf{S}^1$ and register the point obtained in the end point table.

   (a) Obtain the intersection point $\mathbf{P}^2$ between E^2 and $\mathbf{S}^1$ and calculate the parameter of $\mathbf{P}^2$ on the surface $\mathbf{S}^1$.

(b) Calculate the parameter of the intersection point $\mathbf{P}^2$ on the surface $\mathbf{S}^2$.

(c) If E^2 is a trimming edge, obtain the intersection point between the three surfaces, i.e., adjacent two surfaces lying on E^2 and the surface $\mathbf{S}^1$, and regard the point obtained as being the exact intersection point between E^2 and $\mathbf{S}^1$. Otherwise, the intersection point obtained in the step 3a is the intersection point.

(d) Register the parameters of the intersection point on each of the surfaces in the end point table.

4. If the end point table is empty, an intersection curve may be created inside the surfaces. In this case, obtain a point on the intersection curve and register it in the end point table.

5. If the end point table is empty, terminate the processing.

6. Determine the direction of tracing of each end point registered in the end point table by scanning the end point table and by defining the trace vector $\mathbf{v}$.

   Suppose the parameter of the end point $\mathbf{P}_i$ on the surface $\mathbf{S}^1$ is u_i^1, v_i^1, and that on the surface $\mathbf{S}^2$ to be r_i^2, s_i^2. Also suppose the intersection line between the tangent planes of $\mathbf{S}^1(u_i^1, v_i^1)$ and $\mathbf{S}^2(r_i^2, s_i^2)$ is l. Then, the direction vector at $\mathbf{P}_i$ of the intersection curve between the two surfaces have the same direction as l (Figure 6.25).

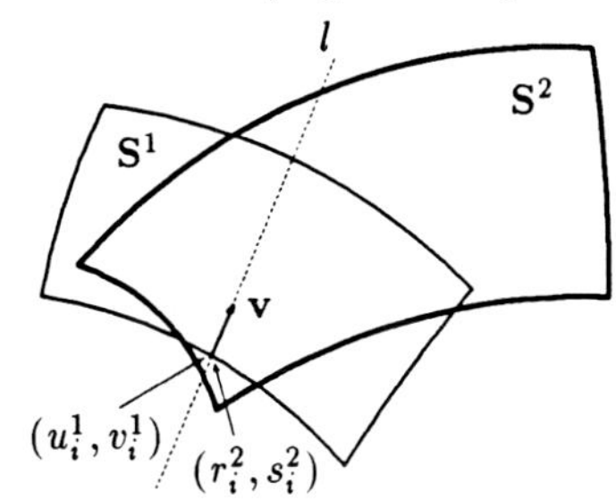

Fig. 6.25. How to determine the direction of tracing

7. Find untraced points in the end point table.

   Obtain the parameters u_i^1, v_i^1 on the surface $\mathbf{S}^1$ and r_i^2, s_i^2 on $\mathbf{S}^2$ of the end point $\mathbf{P}_i$ (i is an index value in the end point table) which trace vector is registered but has not yet been traced. Also obtain the trace vector $\mathbf{v}$.

8. Obtain the points on the intersection curve and the direction vector at each stage of tracing.

   Trace until either of the following conditions is satisfied:

- When the distance between the end point $\mathbf{P}_e$ of the trace vector and $\mathbf{P}_t$ in the end point table is shorter than the length of the trace vector (compare their coordinates).
- The tip of the trace vector lies outside the parametric space of two surfaces.

Set a trace vector of proper length at $\mathbf{P}_i$ in the direction of $\mathbf{v}$ and refine the end point of the trace vector to obtain a point that lies on two surfaces by using the geometric Newton-Raphson method. Then, make the point obtained the new starting point of tracing. By repeating this process until one of the conditions in the step 8 is satisfied, the sequence of end points on the intersection point between the two surfaces can be obtained.

The greater the change in the direction of the trace vector or the closer the direction of the normal vectors of two surfaces, then the shorter the trace vector needs to be defined.

9. Create a sequence of curves from the obtained points and the direction vectors.

10. If two or more sequences of curves are obtained in the step 9, check whether the sequences intersect. If they do, divide the sequence at the intersection point (see Figure 6.26).

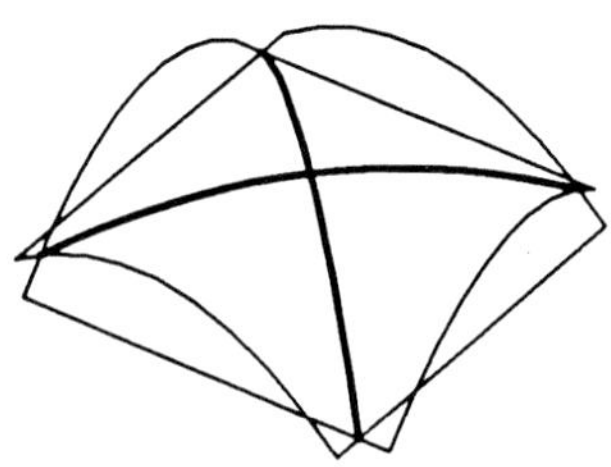

Fig. 6.26. Intersecting intersection curves

In the case of Boolean operations, the curve-surface intersection and surface-surface intersection can be performed separately. An edge usually has different surfaces on either of its sides, and the operation can be made faster if done as follows:

1. Calculate the intersection between all edges of solid A and all surfaces of solid B, and store the relationship.

2. Calculate the intersection between all edges of solid B and all surfaces of solid A, and store the relationship.

The following shows examples modeled in the solid modeler DESIGNBASE, using the methods described so far. In each of Figures 6.27 and 6.28, (a) shows the original two solids and (b) shows the solid created after the Boolean union operation.

Figure 6.27 shows a cylinder and a torus combined. The surfaces of the two solids are represented by quadratic rational Bézier surfaces and the intersection curves are obtained by the marching method.

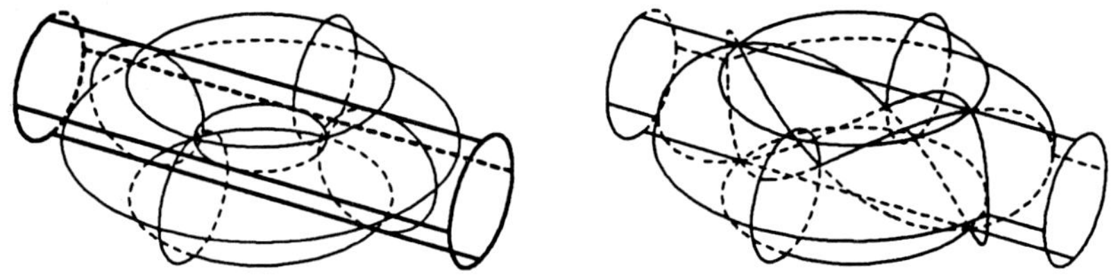

(a) (b)

Fig. 6.27. An example of a union operation of solids containing curved surfaces (1)

Figure 6.28 shows solids with free-form surfaces in a complicated intersection. In this example, the intersection curves intersect themselves, and so the intersection points between them have to be obtained and then subsequently subdivided.

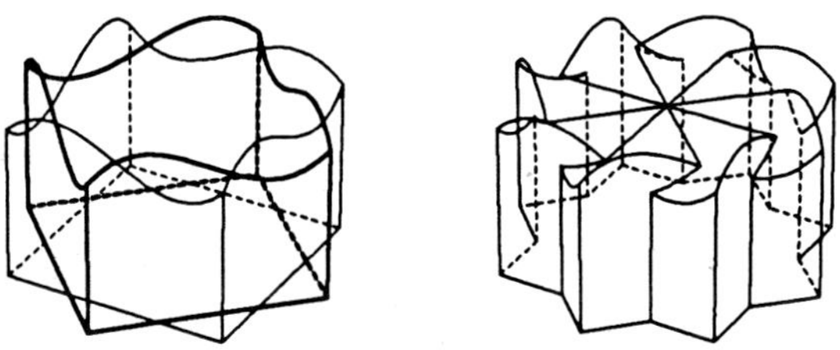

(a) (b)

Fig. 6.28. An example of a union operation of solids containing curved surfaces (2)

6.5 Intersection Calculation in DESIGNBASE

This chapter has explained the algorithms of various intersection calculations. Those algorithms used in DESIGNBASE are listed in Table 6.1. Depending on the elements that are to be intersected, DESIGNBASE can be the most stable and the fastest method for each of the intersection calculations. The Boolean operations described in Chapter 8 are realized by using these intersection calculation routines.

Table 6.1. Intersection calculation algorithms used in DESIGNBASE

curve/curve	straight line/straight line	A-method
	straight line/arc	A-method
	arc/arc	A-method
	free-form curve/straight line	A-method, GNR-method
	free-form curve/arc	GNR-method
	free-form curve/free-form curve	GNR-method
curve/surface	straight line/plane	A-method
	straight line/quadratic surface	A-method
	arc/quadratic surface	A-method
	free-form curve/quadratic surface	GNR-method
	all types of curves/free-form surface	GNR-method
surface/surface	plane/plane	A-method
	plane/ quadratic surface	G-method
	quadratic surface/quadratic surface	G-method
	free-form surface/plane	M-method
	free-form surface/quadratic surface	M-method
	free-form surface/free-form surface	M-method

A-method : Algebraic method
GNR-method : Geometric Newton-Raphson method
G-method : Geometric method
M-method : Marching method

7. LOCAL OPERATIONS

7.1 Introduction

Solid modeling tool-kit as a kernel of 3-dimensional CAD systems should incorporate functions that enable designers to create and modify shapes as they imagine. In particular, future solid modelers should be able to perform flexible modifications to models containing curved surfaces. To freely design models containing curved surfaces, many kinds of modification operations have to be supported in a system. Modification operations can be roughly divided into the following four operations: primitive solid generation, local modification, Boolean operations, and rounding. The local modification is the operation that creates, modifies, or deletes vertices, edges, or loops of a model.

This chapter presents firstly, the methods used to create and modify models containing curved surfaces. The actual modification of models can be performed by combining primitive operations as explained in Chapter 3. Next, the chapter explains how models are created or modified in DESIGNBASE. The Boolean operations are explained in Chapter 8 and the rounding operation is explained in Chapter 9.

7.2 Creation and Modification of Solids with Surfaces

When creating and modifying a model containing free-form surfaces, there are several methods chosen depending on the requirements of how the surface should be formed or on the data to represent the surface. This section explains the methods used to define a surface and provides examples of the four methods: **skinning**, **control point translation**, **surface interpolation**, and **Boolean operations**.

7.2.1 Defining surfaces by skinning

Skinning is a method used to create a model by specifying the cross-section curves of the model to be designed and connecting the curves smoothly. The skinning method has been very popular in defining models containing surfaces. This is because, when making a blueprint of an object containing surfaces, the cross sections of the object are usually drawn, and also it is quite natural to use cross-section data when modeling an object in a computer from the blueprint. In designing marine vessels or aircrafts, their shapes are defined by the cross-section data. Hence skinning is indispensable in defining shapes from shape data. Figure 7.1 shows an example of the skinning method.

(a) (b)

Fig. 7.1. Skinning

7.2.2 Defining surfaces by translating control points

Bézier surfaces or B-spline surfaces are defined by control points and the shape of these surfaces changes by moving the position of their control points. In the control point translation method a surface is modified by changing the position of the control points. As explained in Chapter 5, a Bézier surface or a B-spline surface follows the shape of the defining polygon net, so that designers can easily estimate how shapes will change if the control points are translated. This makes the control point translation quite popular. Figure 7.2(a) shows a Bézier surface and its defining polygon and Figure 7.2(b) shows the shape changed by translating its control points.

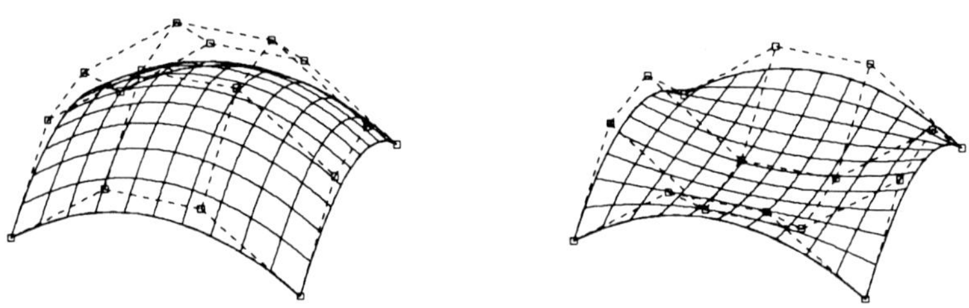

(a) (b)

Fig. 7.2. Control point translation

7.2.3 Defining surface by interpolation

To create a surface by interpolation, a designer needs to define the boundary curves of a surface shape. Some solid modelers or surface modelers incorporate the function to create a surface in this way by using Coons or Gregory patches as the interpolating surfaces. Figure 7.3 shows an example of an interpolation with Gregory patches. When a designer defines the boundaries of a surface as shown in Figure 7.3(a), a system automatically interpolates the surface. Figure 7.3(b) shows the contour lines of the interpolated surface.

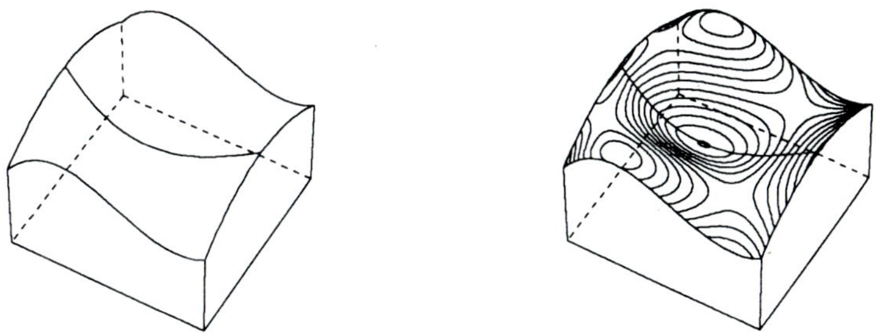

(a) (b)

Fig. 7.3. Surface interpolation

To change the surface shape using this method, the boundary edges must be modified. If a satisfactory shape cannot be obtained by doing this, another boundary edge should be added so that it is possible to have control of the shape. In Figure 7.3, for example, one edge is added at the center of the top surface so that the surface shape is modified, and the surface is pitted.

7.2.4 Defining surfaces by Boolean operations

Boolean operations are indispensable when creating complicated shapes. Although the surfaces of a model are simple, by using the Boolean operations to combine surfaces, a complicated shape can be created. Figure 7.4 shows an example of the Boolean operation. In this figure, cylinders are combined several times to form a complicated shape. Boolean operations are often applied to models which have been created by skinning, control point translation, or surface interpolation.

7.3 Primitive Solid Generation

This section briefly explains how simple models, such as a parallelepiped, rotational model, mirror-image model, sweep model, or skinning model, are created in DESIGNBASE.

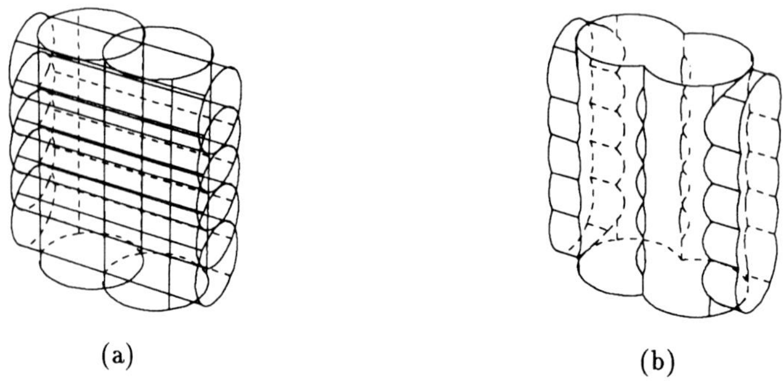

Fig. 7.4. Boolean operation

7.3.1 Creation of a parallelepiped and a cylinder

A parallelepiped is created by defining a rectangular **lamina** and lifting it so as to add thickness to the lamina. The actual primitive operations are as follows:

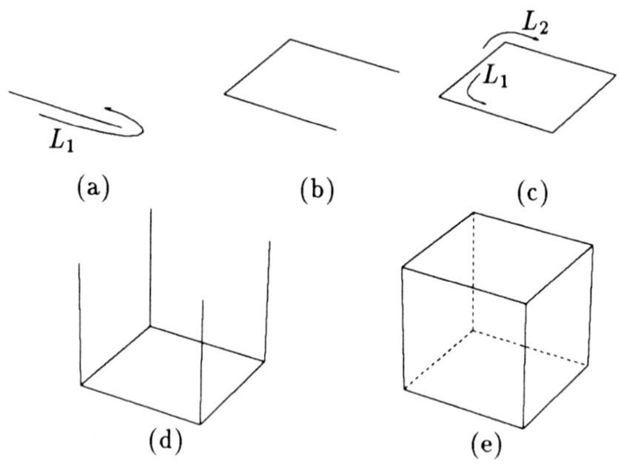

Fig. 7.5. Creation of a parallelepiped

1. Use the primitive operation MEVVL, to create a model that consists of only one edge (Figure 7.5(a)). This edge has a loop L_1

2. Use the primitive operation MEV, to create edges one by one (Figure 7.5(b)).

3. Use the primitive operation MEL, to close the loop, thus forming a rectangle and defining a lamina (Figure 7.5(c)). The loop L_1 is thereby divided into two loops L_1 and L_2. These loops have the same shape but their normal vectors have opposite directions.

4. Use the primitive operation MEV, to create four edges at the vertices of the rectangular lamina in the upper direction (Figure 7.5(d)).

5. Repeat the primitive operation MEL four times, to create four edges so that a parallelepiped is formed (Figure 7.5(e)).

A cylinder can be created in the same way as a parallelepiped. However, two things differ: a circular lamina has to be created instead of a rectangular lamina and data relating to a cylinder should be added to the surface created. In order to create circular shapes, arc shapes must be specified in the the primitive operations MEVVL, MEV, and MEL where these operations create edges.

7.3.2 Creation of a rotational model

A model created by rotating curves on a plane about an axis on the plane is called a **rotational model**. This subsection explains how to create the rotational model by rotating curves defined in the XY-plane about the z-axis as follows:

1. The curve that rotates about the z-axis is called a **definition curve**. Distribute the definition curve every 90° about the z-axis (Figure 7.6(a)).

2. Connect corresponding vertices of adjacent definition curves by using the primitive operations MEKL and MEL (Figure 7.6(b)). The edges created here should be 1/4 arcs.

3. The surface of a rotational model, excluding the top and the bottom, is called the **rotational surface**. Add the surface geometric data to the rotational surface by using a primitive operation MSF to represent the rotational model exactly.

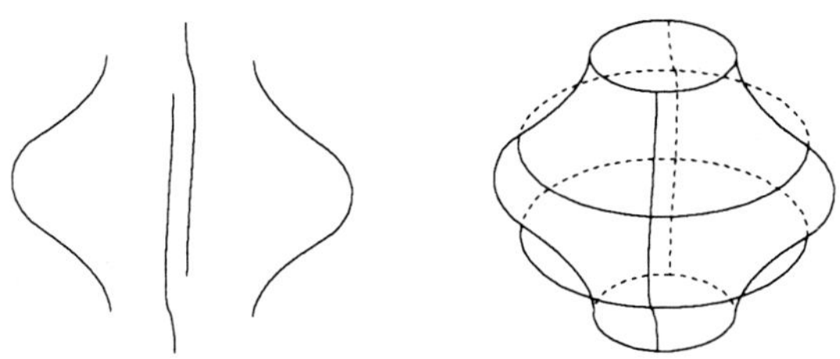

(a) (b)

Fig. 7.6. Creation of a rotational model

In step 3, surface geometric data must be added to the rotational surface because of the following reason: immediately after the edges of the model are created, the surface is automatically interpolated. The interpolated surface may be a little deformed since it satisfies the condition of tangent plane continuity. Therefore, in order to make the surface form a correct rotational model, rational Bézier surface data is added to it[110].

7.3.3 Creation of a mirror-image model

Generally, lots of industrial products are symmetrical in shapes. When such shapes are to be created, it is sufficient to model only a half of the shape. The other half is created as a **mirror-image model**.

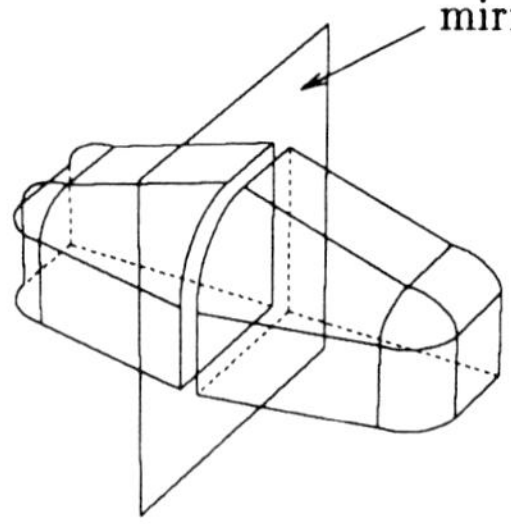

Fig. 7.7. A mirror-image model

As Figure 7.7 illustrates, a mirror-image model is created by defining a plane as the mirror surface of a model and creating its identical model about the plane. The procedure is as follows:

1. Copy the original model and make a new one.
2. Move all the geometric data of the new model such as vertices or control points to the symmetrical position about the plane defined as the mirror surface.
3. Reverse the data of the new model by the primitive operation NEGATE.

7.3.4 Creation of a skinning model

The **skinning model** is created by specifying the cross-section curves of an object and then connecting them smoothly. The procedure is as follows[144]:

1. Create the cross-section curves according to the data specified by using primitive operations MEVVL, MEV, and MEL (see Figure 7.8(a)). This step is the same as that in the creation of the rectangular lamina of a parallelepiped. The cross-sections are created as laminas, made from closed loops.

2. Connect vertices of adjacent laminas by using primitive operations MEKL and MEL (Figure 7.8(b)). The edges should be smooth curves that pass through the corresponding vertices of neighboring laminas.

(a) (b)

Fig. 7.8. Creation of a skinning model

By interpolating the surface created in the above procedure, a model whose cross-section curves are smoothly connected, can be obtained. In the operation, cross-section curves are created as laminas, on condition that the number of vertices of each lamina is equal. If the laminas have different numbers of vertices, their edges should be properly divided so that the number of vertices is equal.

7.3.5 Creation of a sweep model

The **sweep model** is created by specifying a cross-section curve of an object and the trajectory curve of the cross-section curve. The sweep model can be defined with the cross-section curve swept along the trajectory. The procedure is as follows[143]:

1. Divide the trajectory curve at several points.
2. Create laminas of the cross-sections and distribute them on the points along the trajectory curve (Figure 7.9(a)). The center of the lamina should coincide with a point and the normal vector of the lamina should lie parallel to the tangent vector of the trajectory curve.
3. As well as creating a skinning model, create edges that connect neighboring laminas (Figure 7.9(b)). The edges created should be smooth curves lying in parallel to the trajectory curve.

By interpolating the surface created in the above procedure, a sweep model can be obtained.

7.4 Local Modification Operations

Local modification operations are the operations to create, modify, and/or delete topological elements such as vertices, edges, or loops. This section focuses on the local modification operations on edges. The lifting operation of

(a) (b)

Fig. 7.9. Creation of a sweep model

topological elements as well as the cutting and gluing operations are also explained. The implementation methods are based on the local modification operations of DESIGNBASE.

7.4.1 Creation and modification of edges

There are four steps involved in creating an edge:

1. Connect two points in space. The primitive operation MEVVL is used (Figure 7.10(a)).
2. Connect a point in space with a vertex. The primitive operation MEV is used (Figure 7.10(b)).
3. Connect two vertices on a common loop. The primitive operation MEL is used. The loop is then divided into two loops (Figure 7.10(c)).
4. Connect two vertices on different loops. The primitive operation MEKL is used (Figure 7.10(d)).

There are also four steps to delete an edge:

1. A vertex of an edge that is not connected to any other edge is called a **lone vertex**. If both end vertices of an edge to be deleted are lone vertices, the edge is deleted by the primitive operation KEVVL (Figure 7.10(a)).
2. If either end vertex of an edge to be deleted is a lone vertex, the edge is deleted by the primitive operation KEV (Figure 7.10(b)).
3. If neither vertex of an edge to be deleted is a lone vertex and loops on both sides of the edge are different, the edge is deleted by the primitive operation KEL (Figure 7.10(c)).
4. If neither vertex of an edge to be deleted is a lone vertex and loops on both sides of the edge are equal, the edge is deleted by the primitive operation KEML (Figure 7.10(d)).

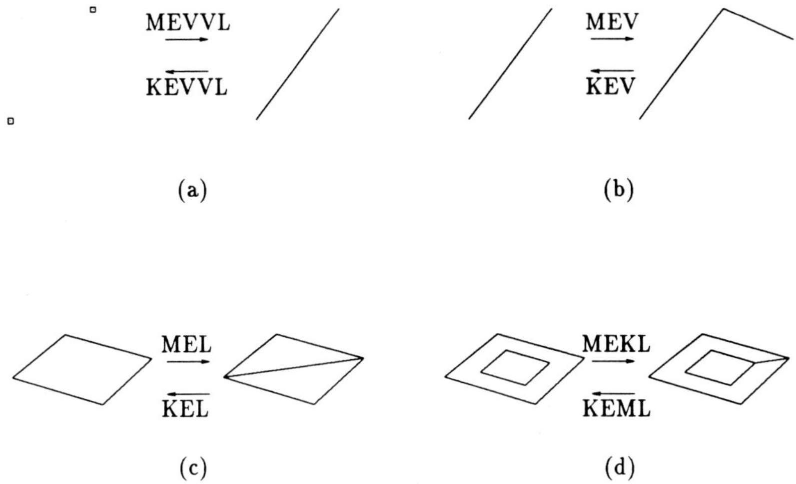

Fig. 7.10. Creation and deletion of an edge

When an edge is to be created on a surface, it must lie on the surface geometrically. For example, the model in Figure 7.11(a) has the surface illustrated in Figure 7.11(d). If another edge is created on the surface as shown in Figure 7.11(b), it does not lie on the surface and the surface shape changes (Figure 7.11(e)). In order that the surface shape remains unchanged when creating an edge on a surface, the edge should lie exactly on a surface. Figure 7.11(c) shows an edge lying exactly on the surface, and the surface shape remains unchanged (Figure 7.11(f)).

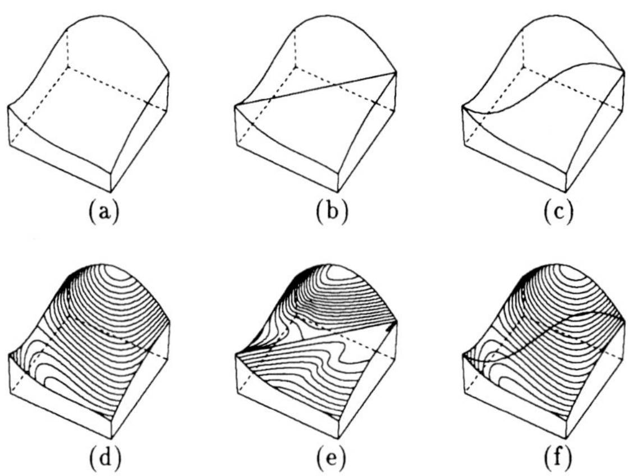

Fig. 7.11. Edges created on a surface

7.4.2 Lifting operation

The **lifting operation** lifts topological elements such as faces, edges, or vertices of a model. Figure 7.12 illustrates an example in which the top face of a parallelepiped is lifted.

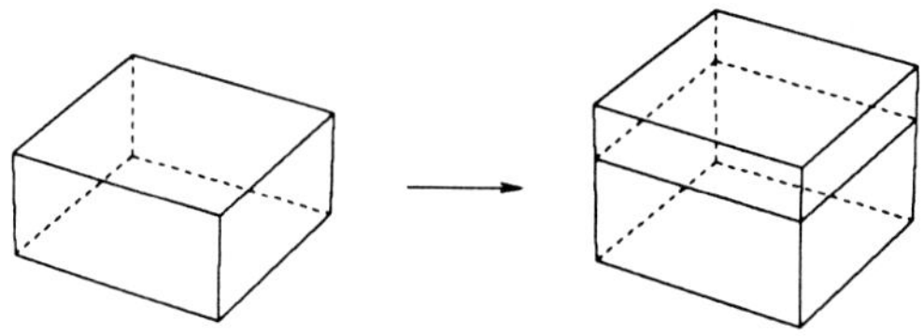

Fig. 7.12. Lifting operation of a face

If a face being lifted contains a C-loop, the C-loop should be lifted at the same time. In Figure 7.13(a), the face F contains a C-loop. The face F can be lifted by the following procedure:

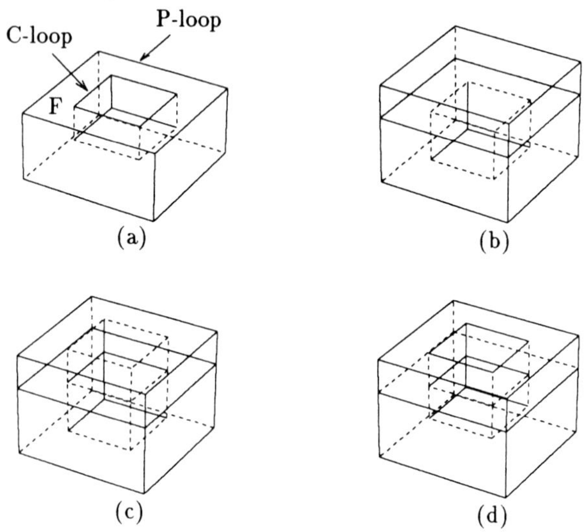

Fig. 7.13. Lifting up a face containing a C-loop

1. By using the primitive operation KCLMPL, change the C-loop into a P-loop.
2. By using the primitive operations MEV and MEL, lift only the original P-loop (Figure 7.13(b)).
3. Lift the original C-loop (Figure 7.13(c)). This step is also done by the primitive operations MEV and MEL.

4. Convert the loop type by using the primitive operation MCLKPL so that the loop lifted in step 3 is changed into the C-loop of the loop lifted in step 2. (Figure 7.13(d)). As a result, the C-loop is lifted at the same time as the P-loop.

7.4.3 Model cutting

Model cutting is an operation to split a model into two parts by a specified plane. The procedure is as follows:

1. Calculate the intersection between a model and a cutter plane (Figure 7.14(a)): first, obtain the intersection points between the edges of the model and the plane. Then obtain the intersection curves between the surface of the model and the plane. Store the data of the intersection points obtained and the curves in a table.

2. While checking the intersection point table, create double vertices on each intersection point by using the primitive operation MVE (Figure 7.14(b)). If a vertex already exists on the intersection point, duplicate the vertex by using the primitive operation MDEV. Between the duplicated intersection vertices, an edge of length 0 exists.

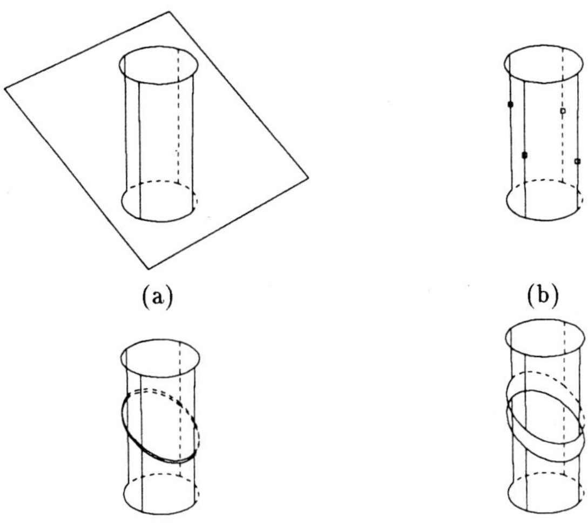

(a) (b) (c) (d)

Fig. 7.14. Model cutting

3. While checking the intersection curve table, create double edges between neighboring vertices by using the primitive operations MEV and MEL (Figure 7.14(c)).

4. Delete the edges of length 0 between duplicated vertices by using the primitive operations KEL and KEML (Figure 7.14(d)).

7.4.4 Model gluing

Model gluing is an operation to combine two touching loops. The gluing operation is performed in two cases: one is the case when one loop completely contains another loop, and the other is when two loops have exactly the same topology and geometry.

Figure 7.15 illustrates the former case. In this case, the gluing operation is completed by changing the type of the loop which is contained in the other loop so that it becomes the C-loop of the containing loop. In Figure 7.15, two loops can be glued only by changing the loop L_1 into the C-loop of the loop L_2.

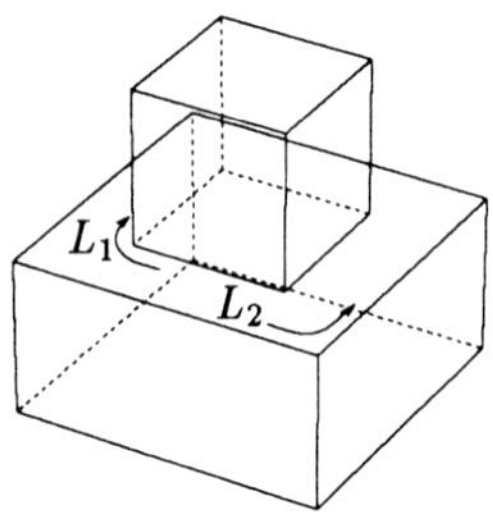

Fig. 7.15. Model gluing (1)

When two loops have exactly the same topology, that is, corresponding vertices have the same coordinates, the gluing operation is performed as follows:

1. By using the primitive operations MEKL and MEL, create an edge of length 0 between corresponding vertices of the two loops (Figure 7.16(b)). The two models are connected on the touching face and junction edges exist on the face.

2. By using the primitive operation KEL, delete one of the junction edges (Figure 7.16(c)).

3. By using the primitive operation KVE, delete the edges of length 0, created in step 1 (Figure 7.16(d)).

This model gluing method can be considered as the reverse operation of model cutting.

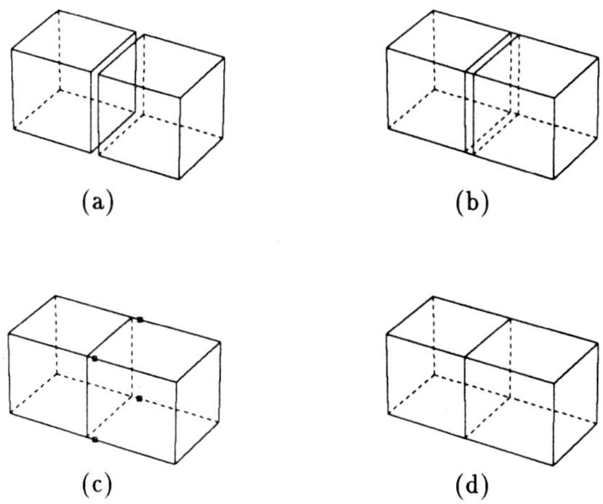

Fig. 7.16. Model gluing (2)

7.5 Programs of Local Operations

This section shows two programs implemented by using the primitive operations explained in Chapter 3.

The first program is to create a parallelepiped with three dimensions x, y, and z. In this program, a parallelepiped is created so that one of its point coincides with the origin of the coordinate system: at first a rectangle is created on the XY-plane, then from each vertex of the rectangle, an edge is created in parallel to the z-axis.

```
/*
 * generateBlock --- Generate a block.
 *
 *  Parameters:
 *    body     (I)  Body ID
 *    x        (I)  Length in the x direction
 *    y        (I)  Length in the y direction
 *    z        (I)  Length in the z direction
 */
generateBlock(body, x, y, z)
    int      body;
    double  x, y, z;
{
    int e1, e2, e3, e4, e5, e6, e7, e8, e9, e10, e11, e12;
    int v1, v2, v3, v4, v5, v6, v7, v8;
    int lp1, lp2, lp3, lp4, lp5, lp6;
    struct dbPoint vect, normal;
    struct dbPoint pnt1, pnt2;
```

```
/*
 * Create a straight line edge which connects two points
 * (0,0,0) and (x,0,0).
 * The ID of the created edge is already set to e1.
 * The point on (0,0,0) is the vertex v1
 * and that on (x, 0, 0) is v2.
 * lp1 is to be set to the ID of the created loop.
 * "clSetCoord" is the macro to set the coordinate
 * to a variable of the "dbPoint" type.
 */
clSetCoord(&pnt1, 0.0, 0.0, 0.0);
clSetCoord(&pnt2, x, 0.0, 0.0);
prMevvl(body, &e1, &v1, &v2, &lp1, &pn1, pnt2);

/*
 * Create an edge to have its both end points on v2 and (x,y,0).
 * The ID of the created vertex is already set to v3.
 */
clSetCoord(&pnt1, x, y, 0.0);
prMev(body, &e2, &v3, &v2, &lp1, &pnt1);

/*
 * Create an edge to have its both end points on v3 and (0,y,0).
 * The ID of the created vertex is already set to v4.
 */
clSetCoord(&pnt1, 0, y, 0.0);
prMev(body, &e3, &v4, &v3, &lp1, &pnt1);

/*
 * Up to here, the following shape is created.
 *
 *      (0,y,0)     (x,y,0)
 *       v4           v3
 *       *------------*
 *            e3      |
 *                    |
 *                    |e2
 *                    |
 *            e1      |
 *       *------------*
 *       v1           v2
 *      (0,0,0)    (x,0,0)
 *
 * Connect v4 and v1.
 */
prMel(body, &e4, &lp2, &lp1, &v4, &v1);

/*
```

```
     * For now, four edges (e1, e2, e3, and e4),
     * four vertices (v1, v2, v3, and v4),
     * and two loops (lp1 and lp2) has been created.
     * Edges grow in the positive direction of the z-axis
     * are to be created next.
     * In the actual implementation of prMel,
     * a new loop lp2 is to be created on the left side
     * of the edge from v4 to v1, and lp2 will have
     * its normal vector in the positive z direction.
     * Create four edges from the four vertices
     * v1, v2, v3, and v4.
     */
    clSetCoord(&pnt1, 0.0, 0.0, z);
    prMev(body, &e5, &v5, &v1, &lp2, &pnt1);

    clSetCoord(&pnt1, x, 0.0, z);
    prMev(body, &e6, &v6, &v2, &lp2, &pnt1);

    clSetCoord(&pnt1, 0.0, y, z);
    prMev(body, &e7, &v7, &v3, &lp2, &pnt1);

    clSetCoord(&pnt1, x, y, z);
    prMev(body, &e8, &v8, &v4, &lp2, &pnt1);

    /*
     * Create edges between four new vertices.
     * Now the creation of a parallelepiped is completed.
     */
    prMel(body, &e9, &lp3, &lp2, &v5, &v6);
    prMel(body, &e10, &lp4, &lp2, &v6, &v7);
    prMel(body, &e11, &lp5, &lp2, &v7, &v8);
    prMel(body, &e12, &lp6, &lp2, &v8, &v5);
}
```

The next program is to delete specified edges. This operation needs to reconstruct the relationship between P- and C-loops. We pay attention to the loop processing here.

```
/*
 * deleteEdge --- Delete a specified edge.
 *
 *  Parameters:
 *    body    (I)  Body ID
 *    edge    (I)  Edge ID
 */
void
deleteEdge(body, edge)
int     body, edge;
{
```

```
int     operation;
int     type1, type2;
int     loop1, loop2;
int     vertex1, vertex2;

/*
 * Obtain the primitive operation to delete an edge.
 * "prGetKillOp" decides the available primitive operation
 * to delete an edge, according to the algorithm
 * described in Subsection 7.4.1.
 */
prGetKillOp(body, edge, &operation, &vertex1);

/*
 * Delete an edge.
 */
switch (operation) {
case PR_KEVVL:
    /*
     * Set the loops on both sides of the edge
     * to loop1 and loop2.
     * loop1 and loop2 are the same in this case.
     */
    rbELoop(body, edge, &loop1, &loop2);
    if (rbLPCType(body, loop1) == DB_CLOOP) {
        /*
         * If loop1 is a C-loop, convert it to a P-loop.
         */
        prKclmpl(body, &loop2, &loop1);
    } else {
        /*
         * If loop1 is a P-loop, convert C-loops of loop1
         * to P-loops by using the function "prFreeCloop".
         */
        prFreeCloop(body, loop1, DB_FALSE);
    }
    prKevvl(body, &edge, &vertex1, &vertex2, &loop1, NULL);
    break;
case PR_KEV:
    /*
     * Delete the edge by using the primitive operation "prKev".
     */
    prKev(body, &edge, &vertex1, &vertex2, &loop1, NULL);
    break;

case PR_KEL:
    /*
     * Check the type of the loops on both sides
```

```
         * of the edge to be deleted.
         */
        rbELoop(body, edge, &loop1, &loop2);
        type1 = rbLPCType(body, loop1);
        type2 = rbLPCType(body, loop2);
        /*
         * To use "prKel", one of the two loops should be
         * a P-loop.  If both loops are C-loops,
         * convert one of the loops to a P-loop.
         */
        if (type1 == DB_CLOOP && type2 == DB_CLOOP) {
            prKclmpl(body, &loop2, &loop1);
        }
        /*
         * Delete the edge by the primitive operation "prKel".
         */
        prKel(body, &edge, &loop1, &loop2, &vertex1, &vertex2, NULL);
        break;
    case PR_KEML:
        /*
         * Delete the edge by the primitive operation "prKeml".
         */
        prKeml(body, &edge, &loop1, &loop2, &vertex1, &vertex2, NULL);

        /*
         * If the created loop loop1 belongs to loop2,
         * define loop1 to be a C-loop of loop2.
         */
        prMakeCloop(body, loop1, loop2);

        /*
         * If the created loop loop2 belongs to loop1,
         * define loop2 to be a C-loop of loop1.
         */
        prMakeCloop(body, loop2, loop1);
        break;
    }
}
```

In the actual program, these functions are generalized to build libraries, and other functionalities can easily be added by using these libraries. In particular, where the processing of curves and surfaces or that of C-loops are troublesome, the modules to deal with such processings are offered to users as general libraries.

8. BOOLEAN OPERATIONS

8.1 Introduction

As described in the previous chapter, Boolean operations are indispensable in the creation of complicated shapes from simple shape models. Unlike local modification operations, the system guarantees that resultant model of the Boolean operations has valid topology and geometry. Because of these characteristics, some B-reps based solid modelers use Boolean operations as the major method for modifying shapes[113,114].

Without considering the problem of numerical errors, the Boolean operations can be performed stably if the models are limited to polyhedrons[90, 115,147,148]. If however the models contain free-form surfaces, then there still remain some problems. These relate to how the intersection calculation between surfaces can be best realized, how the obtained intersection curves should be represented, or how the topological structure of a model should be constructed. This chapter addresses the problems on the Boolean operations of models containing free-form surfaces and explains how those problems are solved in DESIGNBASE.

8.2 Boolean Operations of Solids with Free-Form Surfaces and Their Problems

In solid modelers like DESIGNBASE, which deal with various kinds of surfaces, how should Boolean operations be realized when models contain free-form surfaces? The particular problem that should be considered here is how to calculate intersection curves between the free-form surfaces and how to deal with the intersection curves obtained.

8.2.1 Intersection calculations

The various methods of calculating the intersection between free-form surfaces were explained in Chapter 6. Of these methods, which is the best for Boolean operations? The most important thing is the fact that Boolean operations can be applied to all kinds of surfaces. Because a solid modeler deals with various kinds of surfaces, the marching method is the best since it calculates the intersection between all kinds of surfaces, once they are parameterized. Quadratic surfaces can also be precisely transformed into rational Bézier surfaces, therefore, by using the marching method intersection can be calculated for generalized surfaces (refer to Chapter 3.4.4).

On the other hand, for the natural quadric surface, the geometric method is the best because the calculation is very fast and precise intersection curves can be obtained.

Depending on the types of surfaces therefore, Boolean operations can be executed with high speed and reliability. In DESIGNBASE, therefore, either the marching method or the geometric method is used to perform fast and reliable Boolean operations.

8.2.2 How to deal with intersection curves

A problem that must be considered when executing Boolean operations between free-form surfaces is the matter of how to deal with the intersection curves that are obtained as a result of the calculation. This is because, the degree of the resultant intersection curve becomes very high when the intersection is calculated precisely, making it more difficult to represent the curve. If an intersection curve of high degree is to be represented in a computer, a lot of memory is necessary for storing the data. Furthermore, the high-degree equation can hardly be handled in application programs. Therefore, representing intersection curves precisely is not practical, and it is better to approximate them.

When an intersection curve is approximated, the intersection edge does not lie exactly on the two intersecting surfaces. If Boolean operations are repeated twice or three times between approximated curves, the numerical errors accumulate and it is difficult to calculate the exact intersection. To solve this problem, several approaches have been examined. They are roughly divided as follows:

1. Repatching
2. Trimmed surface in parametric space
3. Three-surface intersection calculation of trimmed surfaces

Let us consider them one by one.

1. **Repatching**

   Repatching is a method used to reconstruct a surface by interpolating it from its boundary edges. (Boundary edges are created on the intersection curves between free-form surfaces[125,149].) With this method, because new surfaces are reconstructed from the boundary curves, intersection curves always lie exactly on the surfaces. Furthermore, by using the repatching method, the surface data is not stored in the computer; hence the amount of data required to represent a solid model can be reduced. Compared to the other two approaches, the amount of data is much smaller.

The repatching, however, has a problem: the surface shape changes after carrying out the Boolean operations. This is because the surface is constructed from the intersection curve data that contains numerical errors, and so it is impossible to create exactly the same shape as the original one. Hence, the shape of the surface changes little by little every time a Boolean operation is performed. The surface shape might even be far from the original one after Boolean operations are executed many times. Moreover, when the boundary curves become very complicated due to the intersection calculation, the surface sometimes cannot be repatched. If so, the surface shape cannot be obtained.

2. **Trimmed surface in parametric space**

After a Boolean operation is executed, a surface is trimmed by intersection. The trimmed surface can be represented by a combination of its shape data and its boundary data. A surface represented with this combination is called a **trimmed surface**. Figure 8.1(a) shows an example of a trimmed surface.

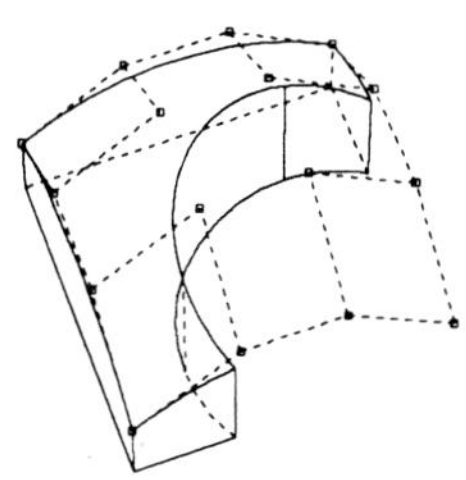

(a)

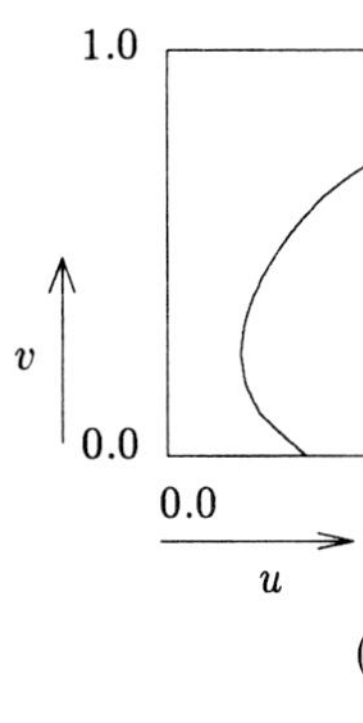

(b)

Fig. 8.1. A trimmed surface in parametric space

When a trimmed surface is used in a Boolean operation, the surface shape does not change after the operation, unlike in the repatching method. After a Boolean operation, all surfaces that intersect with other surfaces are completely represented as trimmed surfaces.

When using a trimmed surface in parametric space, an intersection curve between two free-form surfaces is represented as a curve in the parametric space of the surfaces[25,26,54]. Most of the parametric surfaces define a 2-dimensional space consisting of parameters u and v. A curve on a surface, for example, can be defined as a 2-dimensional curve using the parameters u and v (see Figure 8.1(b)). With this method, the intersection curve obtained after a Boolean operation always lies exactly on a surface. Note that surfaces used in this method should be represented parametrically.

The problem with this method is that the intersection curve, which should be one curve, is represented as a junction curve. This problem occurs because intersection curves are approximated; an exact intersection curve that lies on both intersecting surfaces at the same time cannot be obtained.

To create an intersection curve which is the edge of a model, the intersection curve, defined in parametric space, should be mapped to a curve in 3-dimensional space. However, it is not easy in 3-dimensional space to represent a curve which is defined in parametric space. Because of this restriction, a curve defined in parametric space is approximated and then represented in 3-dimensional space. Hence two intersection curves, slightly different from each other, are created.

3. **Three-surface intersection calculation of trimmed surfaces**

In the method that uses trimmed surfaces, all boundary curves of a surface are represented in 3-dimensional space and are never defined in the parametric space of the surface. With this method, the intersection curve can be uniquely represented. This method can also be applied to surfaces that cannot be represented parametrically. However, the obtained intersection curve does not always lie exactly on the intersecting surfaces. To solve this problem, the **three-surface intersection calculation method** has been proposed[?].

In the model in Figure 8.2(a), for example, the edge E is created as the intersection curve between two trimmed surfaces F_1 and F_2. However, the intersection edge E is not precise. It contains errors since there is a slight difference between E and the exact intersection curve. When this model is intersecting with another model as shown in Figure 8.2(b), the intersection point between the intersection edge E and a surface F_3 has to be calculated. However, the intersection edge E already has errors in itself and thus when the intersection point with the surface F_3 is obtained it will be located at a position different from the exact intersection point. This is a big problem when calculating intersection.

However, because the exact intersection curve can be calculated between the two surfaces F_1 and F_2, the exact intersection point in this example is the intersection point between the three surfaces: the two surfaces on either side of the intersection edge and the surface intersecting with

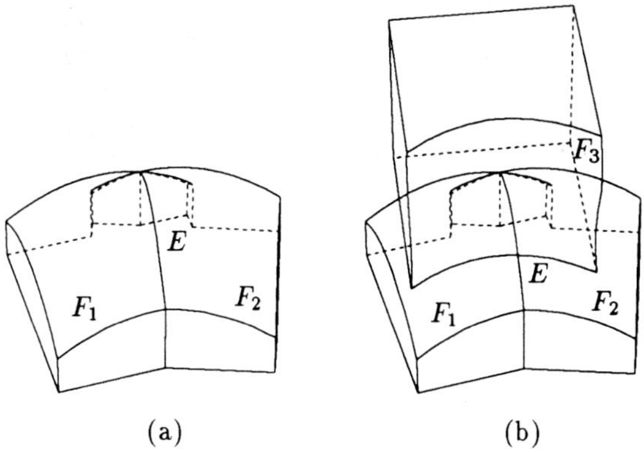

Fig. 8.2. Three-surface intersection point calculation

these two surfaces. In Figure 8.2(b), the intersection point to be obtained is the point where the three surfaces F_1, F_2, and F_3 intersect. Therefore, the intersection point between the exact intersection curve and the surface F_3 can be obtained by calculating the intersection point between the three surfaces F_1, F_2, and F_3. This method of calculating the intersection is called the three-surface intersection calculation. With this method, although an intersection curve is approximated, by referring to the data on the surfaces on either side of the intersection curve, the exact intersection point can be obtained.

8.3 Boolean Operation Algorithm in DESIGNBASE

There are other problems associated with the Boolean operations besides calculating intersection between surfaces. This section explains the detail algorithm of Boolean operations which can be used in practical environment.

As explained before, Boolean operations consist of three operations: union, difference, and intersection operations. The difference and intersection operations can be realized, according to de Morgan's law, by combining the union operation and the operation for reversing the design of a model.

By defining the operation for reversing shapes as the symbol $\neg$, the relationship between the union operation and the difference and intersection operations are indicated as follows:

- difference$(A, B) = \neg(\text{union}(\neg A, B))$
- intersection$(A, B) = \neg(\text{union}(\neg A, \neg B))$

where A and B indicate models. According to the above relationships, the Boolean operations can always be realized using only the union operation.

The following six steps explains the Boolean operations to obtain the union of two solids A and B:

1. Calculate the intersection between edges and surfaces: Check the intersection between all edges and surfaces of two models. If they intersect, calculate the intersection points (see Figure 8.3(a)). If an edge contains numerical errors, calculate the intersection point on the exact intersection curve by using the three-surface intersection calculation method.

2. Calculate the intersection between surfaces: Check the intersection between all surfaces of two models. If they intersect, calculate the intersection curves (Figure 8.3(b)).

3. Create intersection vertices: Create a vertex at every intersection point obtained in the step 1 (Figure 8.3(c)).

4. Create intersection edges: Create intersection edges on the intersection curves obtained in the step 2 (Figure 8.3(d)). If the intersection edge created contains numerical errors, then give the flag that would indicate that the edge is approximated.

5. Detect and delete unnecessary parts: Check both models for parts which intersect and enter into the other and delete the redundant parts (Figure 8.3(e)).

6. Glue the two models: Glue the two models and make them one model (Figure 8.3(f)).

In the following subsections, we shall consider each step in more detail

8.3.1 Intersection calculations between edges and surfaces

The intersection between all edges and surfaces are calculated and the data of the intersection points obtained are stored in the intersection point table. If the edges of the solids are approximated, then the exact intersection points are calculated by using the three-surface intersection calculation method. These points are also stored in the intersection point table. The intersection point table helps to retrieve the coordinates of the intersection points and the information regarding which edges intersect with which surfaces.

The three-surface intersection calculation is realized with the geometric Newton-Raphson method. The procedure is as follows:

1. Suppose three surfaces S_1, S_2, and S_3 are intersecting and the intersection point between an intersection edge of surfaces S_1 and S_2 and surface S_3 is P_0.

2. Project P_0 onto each of surfaces S_1, S_2, and S_3. The projected points are defined as P_1, P_2, and P_3.

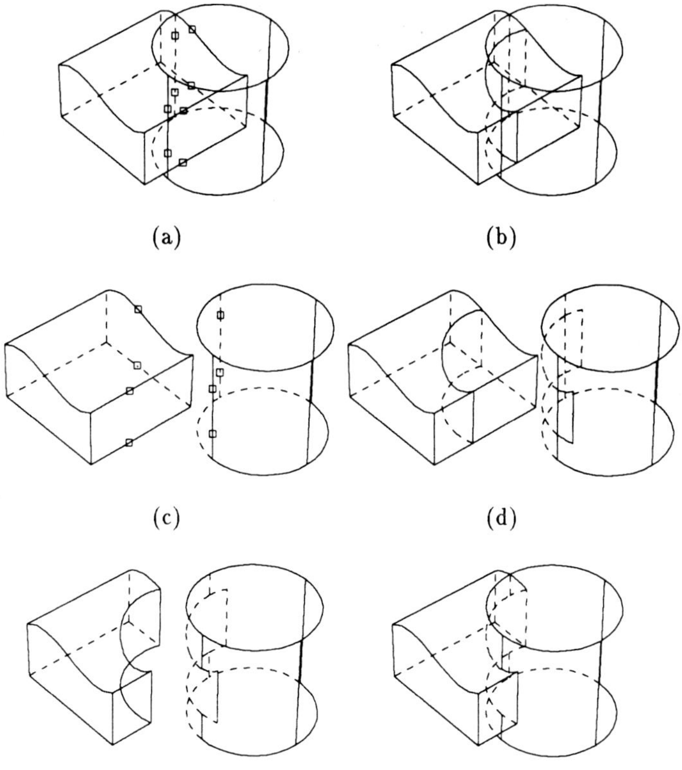

(a) (b) (c) (d) (e) (f)

Fig. 8.3. Boolean operation algorithm

3. If the distance between P_0 and each of P_1, P_2, and P_3 is within the tolerance ε, define P_0 as the intersection point and terminate the calculation.
4. Create a plane F_1 which is tangential to S_1 and passes through P_1. In the same way create planes F_2 and F_3 from P_2 and P_3.
5. Obtain the intersection point P between three planes F_1, F_2, and F_3.
6. Regard $P_0 = P$ and go back to the step 2.

In order to perform this procedure for every approximated intersection edge, DESIGNBASE incorporates a flag — indicating that the intersection edge is approximated — in the fourth step of the Boolean operations.

8.3.2 Intersection calculations between surfaces

To avoid wasteful calculation, bounding boxes of surfaces are created in this step. A bounding box is a minimal polyhedron net that completely covers a surface. With the bounding boxes created, a rough check is performed to examine whether the surfaces intersect. If they do, the intersection curves of the two surfaces is calculated.

To begin, all intersection points necessary for the intersection calculation between the surfaces F_A and F_B should be chosen from the intersection points obtained in the previous step. All the intersection points on the boundary edges of F_A should be retrieved from the table. (The intersection point data includes the information on surfaces with which the edges intersect, and therefore the intersection points on the surface F_B are the necessary intersection points.) Similarly the necessary intersection points about F_B should also be retrieved.

Then, the intersection curves are calculated by using the data from the intersection points obtained and the geometric data on the surfaces. Even if there are no intersection points between the edges and the surfaces, as shown in Figure 8.4, there may be intersection inside of the surfaces. For this reason, the intersections should always be thoroughly checked.

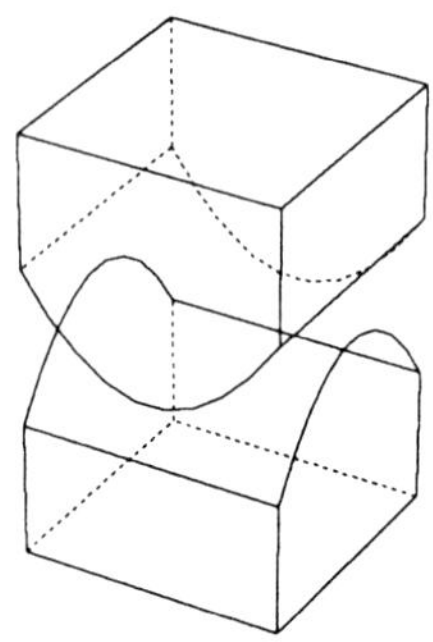

Fig. 8.4. Intersection within surfaces

As stated before, it is difficult to represent an intersection curve precisely. Thus, an intersection curve is approximated with the sequence of curves as listed below:

- Straight line
- Arc
- Cubic Bézier curve
- Quadratic rational Bézier curve
- Cubic rational Bézier curve

An intersection curve can be represented by a sequence of these curves, where the following conditions are satisfied:

- The distance between a surface and both end points of the curve and is within ε.
- The distance between a surface and an arbitrary point on the curve except both end points is within δ.

ε is a constant which validates that two points are the same if they lie within this value. δ is also a constant, where satisfies $\varepsilon < \delta$. In DESIGNBASE, $\varepsilon = 10^{-6}$ and $\delta = 10^{-3}$. δ, the tolerance between curves and a surface, is set relatively larger than ε so as to reduce the costs associated with intersection calculation. If the distance between an intersection curve and both its intersecting surfaces is within ε at any point, the intersection curve can be regarded as exact. On the contrary, if there is a point whose distance to the surface exceeds ε on an intersection curve, the curve is regarded as having errors although the distance between each curve and the surface is within δ.

8.3.3 Creation of intersection vertices

In this step, a vertex should be created at the intersection point obtained from the intersection calculation between an edge and a surface. A vertex is created by the primitive operation MVE. If a vertex to be created lies on an intersection edge containing errors, two intersection points will be obtained: one point between the edge and the surface and the other is the exact intersection point obtained by the three-surface intersection calculation. If there are two intersection points, a vertex is created on the exact intersection point in the following way:

1. By using the primitive operation MVE, create a vertex at the coordinates obtained as the intersection point between an edge and a surface (see Figure 8.5(a)).
2. Calculate the curve data on both sides of the vertex created so that they pass through the exact intersection point and satisfy the tangent continuity condition at that point. If necessary, the curves can be converted into cubic Bézier or cubic rational Bézier curves, before moving the control points.
3. Delete the curve data on the curves of the edges on bith sides of the vertex created in step 1, by using the primitive operation KCV (Figure 8.5(b)).
4. Move the vertex created to the exact intersection point, by using the primitive operation TV (Figure 8.5(c)).
5. Transfer the curve data (obtained in step 2) to the edges on either side of the vertex, by using the primitive operation MCV (Figure 8.5(d)).

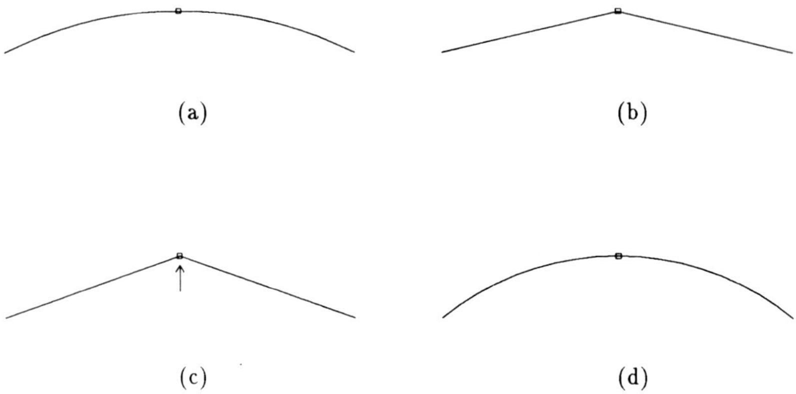

Fig. 8.5. Vertex created on an edge containing errors

8.3.4 Creation of intersection edges

Intersection edges are created by using the primitive operations MEV, MEL, MEVVL, MEKL, and KCLMPL. If an edge is created that contains a certain tolerance, a flag should be given that indicates this; the tolerance about the edge is δ and within δ the edge is regarded as a correct intersection edge.

8.3.5 Classification of boundaries and deletion of unnecessary parts

At the end of the previous step, intersection edges are created on both solid models. In this stage, we can find that each solid is divided into several areas by the intersection edges created. This step detects which areas of one solid lie inside of the other and deletes vertices, edges, and surfaces of the overlapping areas.

The surface of a solid is divided into several areas by the intersection edges. These area are classified as follows:

1. Area that lies inside of the other solid

2. Area that lies on the surface of the other solid

3. Area that lies outside of the other solid

Every surface of a solid belongs to one of the above three types. Unnecessary areas mean all areas that lie inside of the other solid and/or overlapping areas that lie on the surface of the other solid. It is important to check the type of area of a surface. Figure 8.6 illustrates an example of a cross section perpendicular to the intersection edge between solids A and B.

There are nine cases of intersection between two surfaces as shown in Figure 8.7. Ultimately these nine cases can be classified into four types. In

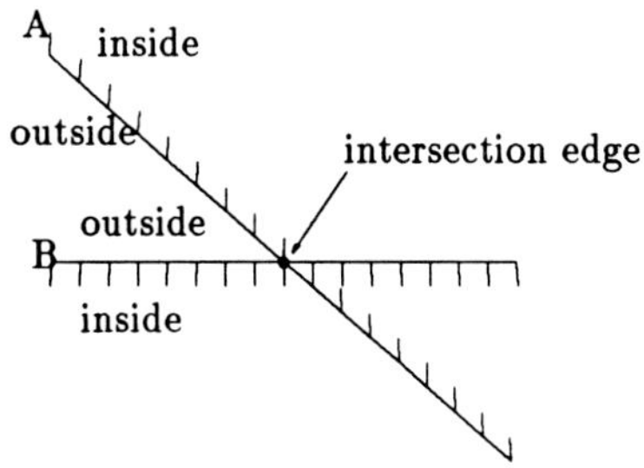

Fig. 8.6. Intersecting surfaces connected to intersection edges

Figure 8.7, the area that lies inside of the other solid is labeled +, the area that lies on the surface of the other solid is 0, and – is the area that lies outside of the other solid. The area labeled 0 is further classified as shown in Table 8.1. Every surface that is adjacent to an intersection edge has a labeling. The area enclosed by a surface labeled – is an unnecessary area. All edges in the unnecessary area should be deleted by primitive operations.

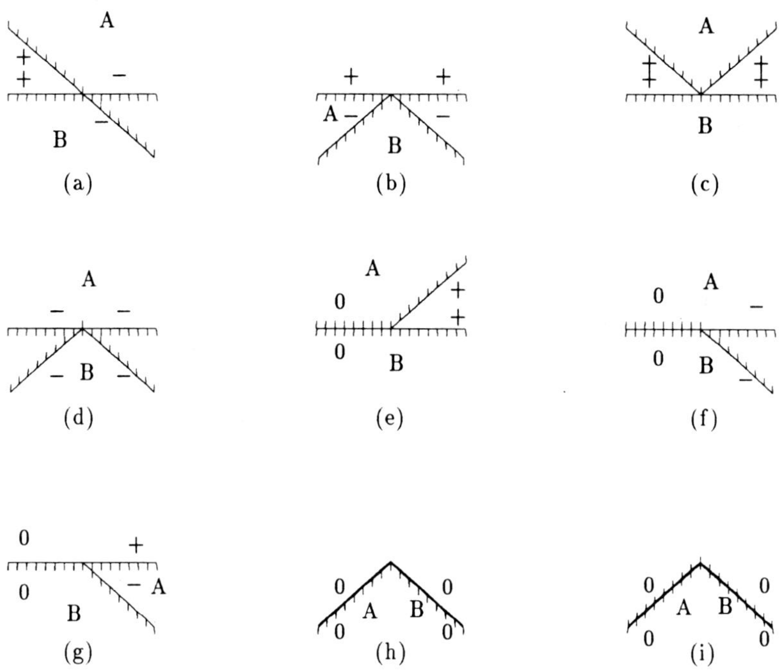

Fig. 8.7. Classification of intersections

Table 8.1. Classification of labels

	SOLID A	SOLID B		SOLID A	SOLID B
case (a)	+−	+−			
case (b)	++	−−			
case (c)	++	++			
case (d)	−−	−−			
case (e)	+0	+0	→	+−	+−
case (f)	−0	−0	→	−−	−−
case (g)	+0	−0	→	++	−−
case (h)	00	00	→	++	++
case (i)	00	00	→	−−	−−

8.3.6 Gluing of two solids

In this last step, two solids are glued. On completion of the previous step, the two solids have no intersecting parts and are touching where they should be glued. The two solids are combined into one solid by applying the gluing operation described in Section 7.2.

8.3.7 Examples of execution

Examples in which the Boolean operations are used are shown in Figure 8.8. Figure 8.8(a) shows a machine part, (b) shows a solid presented in the modeling exercises of CAM-I, and (c) shows a vacuum cleaner.

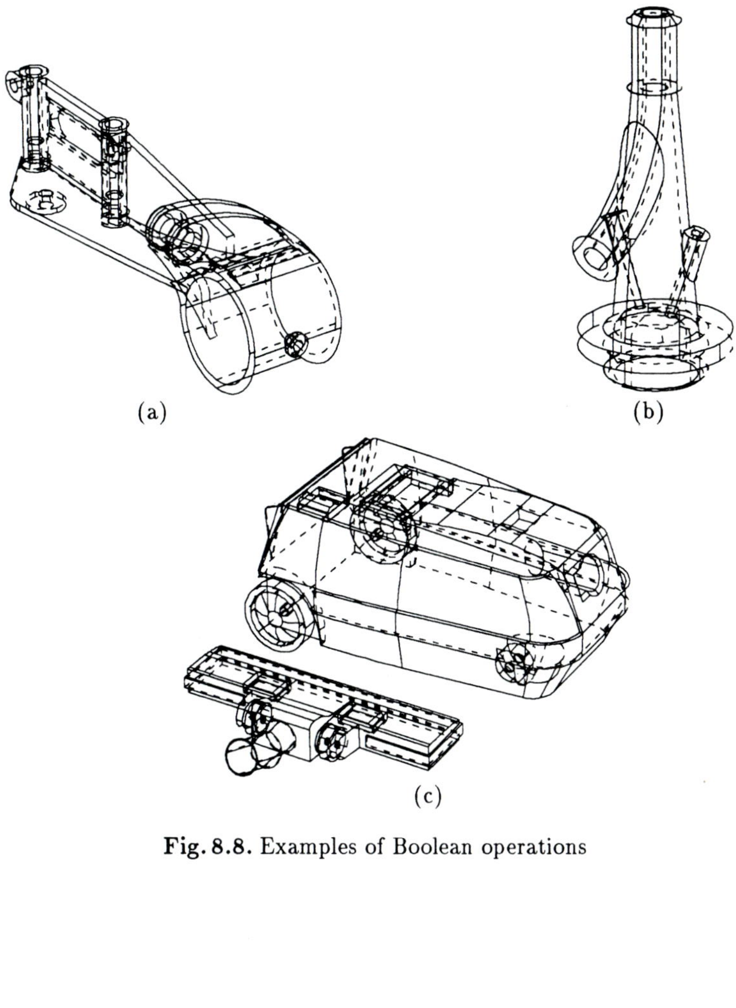

Fig. 8.8. Examples of Boolean operations

9. ROUNDING OPERATION

9.1 Introduction

Most solids modeled by a CAD system can generate **fillet surfaces** which connect sharp corners smoothly. The fillet surfaces help to reinforce machine parts, to make processing easy, and to provide aesthetic shapes. The **rounding operation** is an operation to create the fillet surfaces. This chapter describes the problems associated with the rounding operation and explains how to input and represent fillet surfaces in DESIGNBASE.

9.2 Problems with the Rounding Operation

To perform the rounding operation in B-reps solid modelers, there are two major problems which must be overcome. One is how the shape of fillet surfaces should be represented by mathematical surface equations and the other is how topological elements such as edges should be created. To represent the shapes of fillet surfaces mathematically, Holmström[66] and Martin[94] suggest analytic surfaces such as cylinders. The use of analytic surfaces in the rounding operation can be realized by limiting the types of surfaces used for creating the fillet surfaces. Figure 9.1 shows the creation of a fillet surface using a cylindrical surface.

(a) (b) (c)

Fig. 9.1. Creation of a fillet surface from an analytic surface

Rossignac[119] and Rockwood[117] have proposed a method of representing the shape of fillet surfaces by using free-form surfaces. Choi[37] and Varady[151] have proposed another method to approximate fillet surfaces by using a trajectory drawn from the tangent points of a ball rolling on two surfaces.

In DESIGNBASE, in addition to the method for expressing fillet surfaces mathematically, a method for representing topological elements that construct fillet surfaces has been developed, so that the rounding operation between curved surfaces has been realized. This method can be applied to all types of surfaces used in DESIGNBASE.

9.3 Rounding Operation in DESIGNBASE

The rounding operation in DESIGNBASE[63] has been designed to overcome the two problems described in the previous section. This operation is an expansion of the rounding operation applied to polyhedrons[33]. The characteristics of the rounding operation are as follows:

- In DESIGNBASE, the rounding operation can be performed between natural quadric surfaces such as cylindrical surfaces and free-form surfaces such as Bézier surfaces (see Figure 9.2).

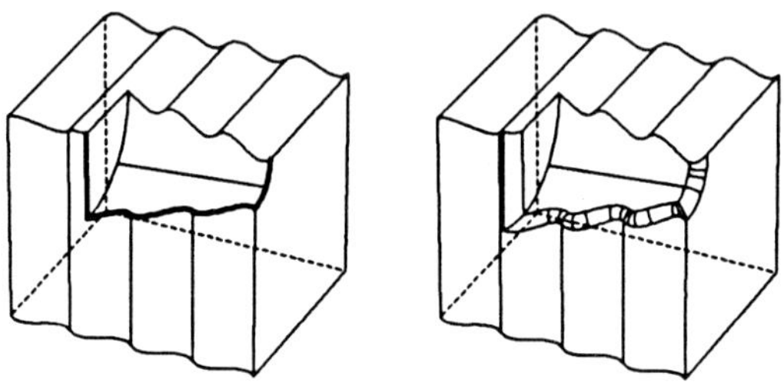

(a) (b)

Fig. 9.2. Rounding operation between curved surfaces

- Gregory patches and rational boundary Gregory patches are introduced to represent the shape of fillet surfaces mathematically. Because of the characteristics of these patches, which are indispensable in the representation of fillet surfaces, two patches can be smoothly connected and non-rectangular irregular meshes can be interpolated (see Figure 9.3).

- Four types of fillet surfaces can be created. The **rolling ball method** is the most popular method for creating these fillet surfaces. This method defines a fillet surface by using the trajectory of a ball rolling on two surfaces. (However, when designing an object, the width of a fillet surface is sometimes required to be constant.) DESIGNBASE supports the four types of fillet surfaces by using the rolling ball method. The four types are explained in detail in the next section.

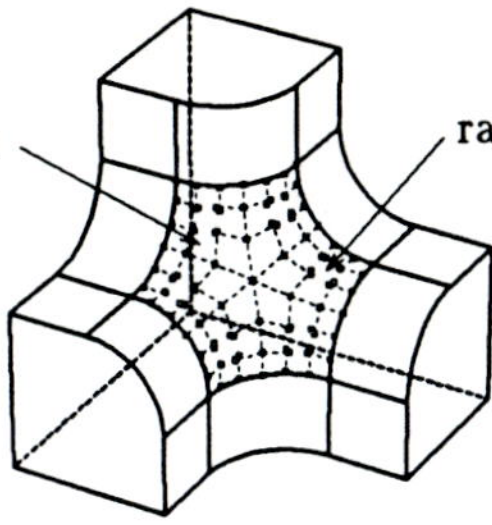

Fig. 9.3. Characteristics of a rational boundary Gregory patch

- **Variable radius blending**, in which the cross-sectional radius gradually changes, can be effected. This method can smoothly connect two fillet surfaces of different cross-sectional radii. In this sense, variable radius blending is the indispensable function in 3-dimensional CAD systems.

9.4 Example of Rounding Operation Execution

The rounding operation in DESIGNBASE is executed in the following way:

1. To edges and vertices of a solid which are to be rounded, specify the **rounding type** and the **rounding radius**.
2. Execute the rounding operation command.
3. If the specified rounding type or rounding radius is required to be modified at the end of the operation, go back to the step 1 by using the UNDO function (refer to Chapter 10) and specify another value as the rounding type or radius.

DESIGNBASE supports four types of fillet surfaces. A fillet surface can be created by specifying a rounding type at an edge. The five rounding types for edges are as follows:

0: An edge is not rounded.

1: An edge is rounded to have the natural shape of a cylinder or cone surface (see Figure 9.4(a)). With this type, the system automatically determines how to round the edge, and an unexpected, novel fillet shape may sometimes be created. Some Japanese computer graphic artists are fond of using this function in DESIGNBASE for doing their work.

2: An edge is rounded by rolling a ball of radius r (Figure 9.4(b)). This is the so-called "rolling ball method", described earlier. The cross-sectional radius of the fillet surface created can be constant, but it varies greatly depending on the angle between the two surfaces connected to the edge to be rounded. Type *3* or *4* is preferred for obtaining smooth change in the cross-sectional radius.

3: An edge is rounded within a distance *d* from either side of the edge (Figure 9.4(c)). The cross section of the fillet surface created will be asymmetrical when two different values are specified to *d*. This rounding type, if used with the variable radius blending function, can create complex fillet shapes, since four different values can be specified for *d* at an edge; two values for right and left *d* at the starting point of the edge and another two at the end point of the edge.

4: An edge is rounded to have a fillet surface of width *w* (Figure 9.4(d)). The width of the fillet surface created can be constant. This rounding type can reduce the variation of the cross-sectional radius.

Of the above five rounding types, the rounding radius is valid in types 2, 3, and 4.

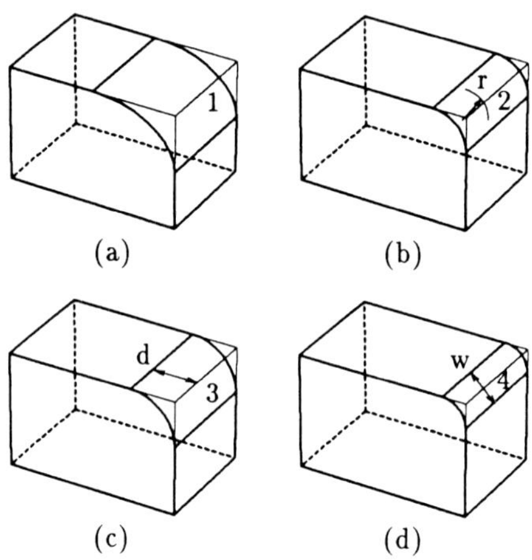

Fig. 9.4. Rounding types for edges

Figure 9.5 shows an actual example of the rounding operation between surfaces in DESIGNBASE. The solid in this figure is created by the Boolean union operations between a cylinder and a cone. The rounding type *2* (rolling ball method) and the rounding radius 1.0 are specified at the edges shown in bold and the operation is executed. As a result, fillet surfaces are created to smoothly connect two quadric surfaces as shown in Figure 9.5(b).

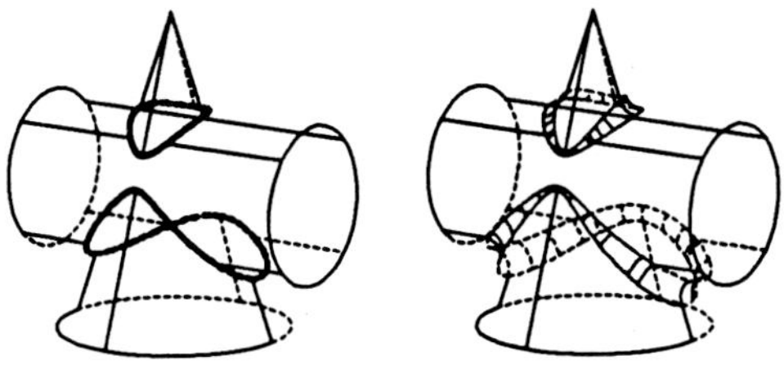
(a) (b)

Fig. 9.5. An example of rounding operation

9.5 Algorithm of Rounding Operation

This section explains the algorithm of the rounding operation in DESIGNBASE. In the following explanation, the edges shown in bold in Figure 9.6(a) are rounded so as to create fillet surfaces. The rounding operation is executed through the five steps as follows:

- Step 1

  Roll a ball along two surfaces connected at an edge to be rounded (shown in bold in Figure 9.6(a)) and calculate the trajectories drawn by the tangent points between the ball and the surfaces (Figure 9.6(b)). The radius of the ball is the cross-sectional radius of the fillet surface to be created. The method used to calculate the trajectory of the tangent points is explained in detail in Section 9.6.

- Step 2

  Create vertices by using the primitive operation MVE at the points where the trajectories intersect with edges of the solid. The edges are divided into several shorter edges by the vertices created (Figure 9.6(c)).

- Step 3

  Create edges along the trajectories to connect the vertices created in the step 2, using the primitive operations MEV and MEL. As a result of this operation, new edges are created on either side of the edge to be rounded (Figure 9.6(d)). An edge created in this step is called an R-edge.

- Step 4

  Delete the edges to be rounded by using the primitive operations KEL and KEV. Next, create arc edges to connect the R-edges by using the primitive operation MEL (Figure 9.6(e)). This process creates the edges that will be the boundaries of the fillet surfaces. The inner shape of a fillet surface is generated by interpolating its boundary edges.

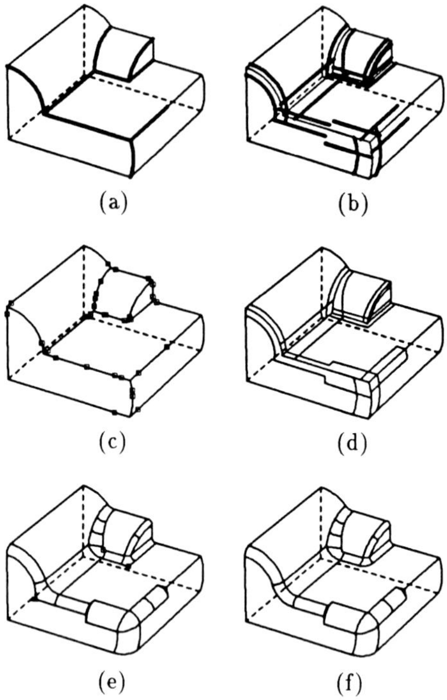

Fig. 9.6. Algorithm of rounding operation

- Step 5

  Delete a vertex that connects two R-edges by using the primitive operation KVE and change the two R-edges into one smooth curved edge (Figure 9.6(f)). This processing creates a free-form surface that smoothly connects the fillet surfaces around the deleted vertex.

The explanation in this section was limited to the procedure for the rolling ball method (rounding type 2). Generally, all the steps, with the exception of step 1 can apply to the other rounding types.

9.6 How to Calculate Trajectories Drawn by Tangent Points

This section explains in detail the method used to calculate trajectories drawn by the tangent points between a ball and two surfaces. The geometric Newton-Raphson method is used to realize this method.

- Step 1

Set the parameter t on an edge to be rounded to 0.0 and set the number of divisions of the edge to n (a natural number). The parameter on an edge is used to calculate the position of a point on the edge. The position where $t = 0.0$ corresponds to the starting point of the edge. The larger the number n, the more precise the trajectories.

- Step 2

  If the parameter t is greater than 1.0, go to step 10. Otherwise, obtain the position $\mathbf{P}_0$ of a point at the parameter t on the edge and r the cross-sectional radius of the fillet surface to be created. Here, r is variable and is a function of t.

- Step 3

  Set the initial points $\mathbf{P}_1$ and $\mathbf{P}_2$ of the geometric Newton-Raphson method on the two free-form surfaces $\mathbf{S}_1$ and $\mathbf{S}_2$ that connect to the edge to be rounded. Set $\mathbf{P}_0$ on the edge using the two initial points $\mathbf{P}_1$ and $\mathbf{P}_2$. After that, calculate the parameter (u_1, v_1) corresponding to the point $\mathbf{P}_1$ on the surface $\mathbf{S}_1$ and the parameter (u_2, v_2) corresponding to the point $\mathbf{P}_2$ on the surface $\mathbf{S}_2$.

- Step 4

  Calculate the tangent plane $\mathbf{F}_1$ at the starting point $\mathbf{P}_1$ on the surface $\mathbf{S}_1$ and $\mathbf{F}_2$ at the other starting point $\mathbf{P}_2$ on $\mathbf{S}_2$ (see Figure 9.7(a)). When the parameter (u, v) on the free-form surface $\mathbf{S}$ is given, a normal vector $\mathbf{n}$ of a tangent plane on the free-form surface can be calculated by the following equation:

$$\mathbf{n}'(u,v) = \frac{\partial \mathbf{S}(u,v)}{\partial u} \times \frac{\partial \mathbf{S}(u,v)}{\partial v}, \quad \mathbf{n} = \frac{\mathbf{n}'}{|\mathbf{n}'|}$$

- Step 5

  Calculate a plane $\mathbf{F}_0$ that passes through the point $\mathbf{P}_0$ and is orthogonal to the two planes $\mathbf{F}_1$ and $\mathbf{F}_2$. The normal vector $\mathbf{n}_0$ of the plane $\mathbf{F}_0$ can be obtained from the outer product of the two vectors $\mathbf{n}_1$ and $\mathbf{n}_2$ of the two planes $\mathbf{F}_1$ and $\mathbf{F}_2$, respectively.

- Step 6

  Fit the ball of radius r obtained in the step 2 on the two planes $\mathbf{F}_1$ and $\mathbf{F}_2$ (Figure 9.7(b)). The center of the ball $\mathbf{C}_0$ should lie on the plane $\mathbf{F}_0$. Then obtain the two tangent point $\mathbf{P}'_1$ and $\mathbf{P}'_2$ between the ball and the two surfaces.

- Step 7

  Check $\mathbf{P}_1$ and $\mathbf{P}'_1$ to see whether they can be regarded as the same point. If they can be regarded as the same, then the point is a tangent point

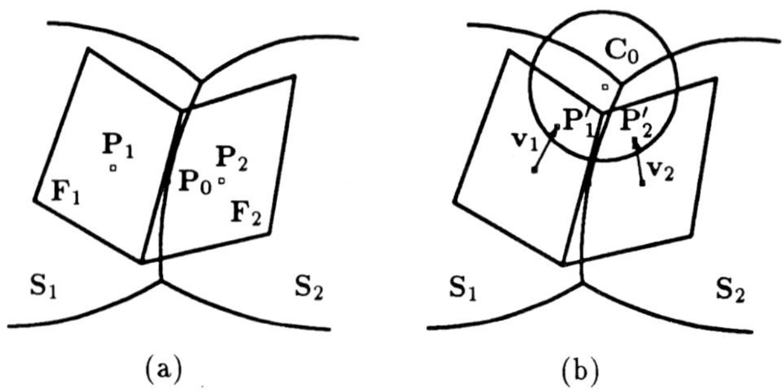

Fig. 9.7. How to calculate trajectories of tangent points between a ball and two surfaces

to the surface $\mathbf{S}_1$. As such $\mathbf{P}_1$ should be stored in an array. Perform the same processing for $\mathbf{P}_2$ and $\mathbf{P}'_2$. Then, add $1/n$ to the parameter t and go back to the step 2.

- Step 8

By using $\mathbf{P}'_1$ and $\mathbf{P}'_2$, obtain the next starting point (u'_1, v'_1) on the surface $\mathbf{S}_1$ and (u'_2, v'_2) on $\mathbf{S}_2$ using the following equations:

$$\begin{aligned}\mathbf{v}_1 = \mathbf{P}'_1 - \mathbf{P}_1 &= \Delta u_1 \frac{\partial \mathbf{S}_1(u_1, v_1)}{\partial u} + \Delta v_1 \frac{\partial \mathbf{S}_1(u_1, v_1)}{\partial v} \\ \mathbf{v}_2 = \mathbf{P}'_2 - \mathbf{P}_2 &= \Delta u_2 \frac{\partial \mathbf{S}_2(u_2, v_2)}{\partial u} + \Delta v_2 \frac{\partial \mathbf{S}_2(u_2, v_2)}{\partial v}\end{aligned}$$

where Δu_1 and Δv_1 can be obtained by calculating two inner products: one between the upper equation and the vector $\partial \mathbf{S}_1(u_1, v_1)/\partial u$ and the other between the upper equation and the vector $\partial \mathbf{S}_1(u_1, v_1)/\partial v$. Then solve the linear simultaneous equation with the two unknowns Δu_1 and Δv_1 and calculate the new start point specified by $u'_1 = u_1 + \Delta u_1$ and $v'_1 = v_1 + \Delta v_1$. The other starting point (u'_2, v'_2) can be obtained in the same way using the lower equation.

- Step 9

Assign (u'_1, v'_1) to (u_1, v_1) and (u'_2, v'_2) to (u_2, v_2). By using these two parameters, calculate the new position of the two points $\mathbf{P}_1$ and $\mathbf{P}_2$, then go back to the step 4.

- Step 10

Retrieve every tangent point $\mathbf{P}_1$ stored in the array in step 7 and create a sequence of curves $\mathbf{T}_1$ that passes through all the points. Create another sequence of curves $\mathbf{T}_2$ from every tangent point $\mathbf{P}_2$ in the array. In this way two sequences of curves — one sequence on one side of the edge to be rounded and the other sequence on the other side — which represent the trajectory of the tangent points are calculated.

This rolling ball method is applied in variable radius blending.

As is described in step 2, r, the cross-sectional radius of an edge, is variable and is the function of the parameter t on the edge. So, for example, to create variable radius blending that varies linearly, the cross-sectional radius r is defined as the linear function of the parameter t. Accordingly, the radius r is varied linearly; r_1 when $t = 0.0$ and r_2 when $t = 1.0$. Complex fillets can also be created by defining r to be a complex function of t.

In designing the shape of a television set, for example, the rounding operation is executed using the above algorithm. In Figure 9.8, the edges shown in bold, indicate where the fillet surfaces are to be created in the television set.

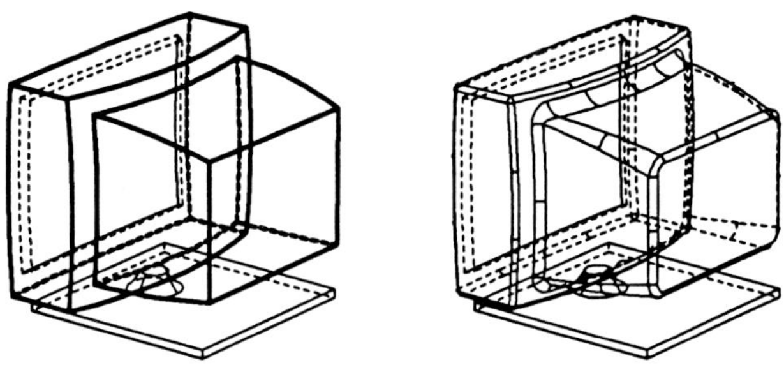

(a) (b)

Fig. 9.8. A TV set model whose edges are rounded

10. FUNCTIONS IN AIDING DESIGN

10.1 Introduction

In the initial stages of designing a product, there will be various requirement regarding the product's appearance, quality, reliability, or cost. Often, it is difficult to satisfy all these requirements completely. Conventional CAD systems enable designers to model the required geometric shape, although this does not always meet with the desired requirements. If it does not, the shape has to be corrected. In addition, the changing requirements of a product means that the product shape has also to be modified sometimes. Because of this, the functions that easily modify the product shape once it is designed are indispensable to design flexibility and productivity. In conventional CAD systems, however, functions that aid design have not been fully supported, thus making the modification in product shape very difficult.

In recent years, new 3-dimensional CAD systems have been developed which support design flexibility and productivity. This chapter explains **parametric design**, a design-aiding function that has gradually become popular in solid modelers and the **UNDO · REDO/ReExecution function**, which assist design in DESIGNBASE.

10.2 Parametric Design

Parametric design is an approach that defines the shape of a product by using the geometric constraints and the dimension data of topological elements. The geometric constraints describe the relation between the design elements, for example, two elements lie in parallel, they are touching, or they are orthogonal to each other. The dimension data can describe, in addition to the dimensional value of elements, the dimensional relation between the elements by using mathematical equations. Parametric design constructs the required shape by solving the equations that express the dimensional relation between the elements, using the geometric constraints.

Parametric design defines the shape of a product by the following steps:

1. Input a 2-dimensional shape as a rough sketch.

2. Input geometric constraints and dimensional data.

3. Reconstruct the 2-dimensional shape by using the specified geometric constraints (see Figure 10.1 (a)).

4. Repeat steps 2 and 3 until the requirements specified are satisfied (Figure 10.1 (b)).

5. Create a 3-dimensional shape by thickening the 2-dimensional shape or rotating it. The values specified in this step for thickness or angles can also be used as dimensional data, thus a shape can easily be modified if necessary.

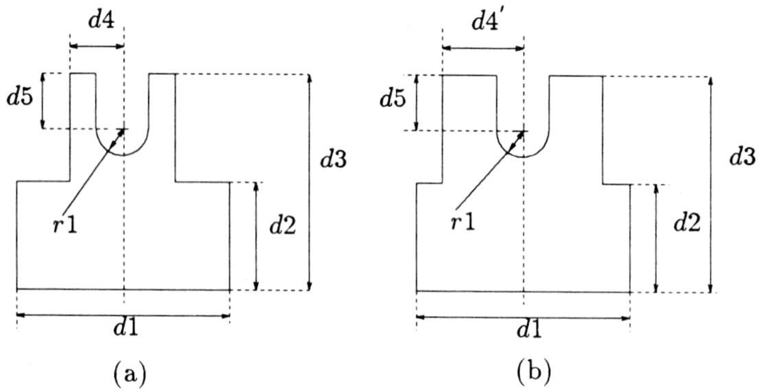

Fig. 10.1. Parametric design

A shape can be modified by only changing the geometric constraints or the parameters of the dimensional data specified. Therefore there is no need to modify the geometric shape directly. As such, a designer is free from having to consider how to modify a shape in order that the desired shape is obtained. He needs only to think about the shape of a product that will satisfy the requirements of the design.

The parametric design can be said to be an excellent function in aiding design because its method of defining shapes is based on the primary intention of a designer. However, parametric design defines a shape by inputting 2-dimensional data, and so it is difficult to model a shape that cannot be easily defined by geometric constraints and dimension data. Moreover, parametric design contains another drawback, that is, when a designer specifies too much geometric constraints and dimension data, it takes a long time to reconstruct the shape and the system will no longer be interactive. Chung[38] concludes that parametric design is insufficient as a design aid because of the limitation on the combination of geometric constraints and dimensional data; he suggests the use of **variational design** which can arbitrarily set the combination.

10.3 UNDO · REDO Operations

10.3.1 Interactive system

When a designer interactively uses a CAD system, the system must be able to deal with the designer's operational errors. The **UNDO** operation, which cancels a previous operation, has been supported in various interactive systems as a restoration function. More recently, this function has been supported as a basic function. In systems used for aiding creative work, however, the UNDO function should be more positively utilized and positioned as the function to aid trial and error in design. This way a designer can repeat operations and redesign many times.

In the usual design process, the last stage of design does not necessarily produce the best shape; the designer chooses the best shape from several previous designs. If the UNDO operation can retrieve lost designs, then trial and error in design can be more powerfully aided. However, it is almost impossible to retrieve a lost design if the UNDO operation is realized by use of the stack operation in restoring data from previous operations. To retrieve any shape in the design process, the system should store the design history in the tree structure, instead of by the stack operation.

The importance of the UNDO operation in an interactive system can be clearly understood, but to fully aid creative work, the **REDO** operation, which retrieves a past design, should also be supported in a system. The system that supports both the UNDO and REDO operations can be said to be the one that truly aids trial and error of design by a computer[55].

10.3.2 UNDO · REDO operations in DESIGNBASE

As an interactive solid modeler, DESIGNBASE supports the UNDO · REDO operations to aid trial and error in design. Any shape designed in a past process can be retrieved with these operations. These operations can be realized in modeling a solid by using primitive operations only. While the primitive operations have their reverse operations, the UNDO · REDO functions ensure that every primitive operation can be restored.

When a solid is modeled using DESIGNBASE, the design history of the solid is stored in the tree structure belonging to the solid. This tree structure is called the **design history tree**. Since the solid is modified by the primitive operations only, what is stored in a design history tree is the primitive operations that have been executed. A design history tree is formed as shown in Figure 10.2.

Figure 10.2 represents the following steps in design:

1. Create a regular octagonal prism.
2. Create a ring on the top surface.
3. Lift the ring.

4. Scale down the lifted face.
5. Perform the UNDO operation.
6. Perform the UNDO operation again.
7. Push down (dig) the ring.
8. Scale down the face pushed down.

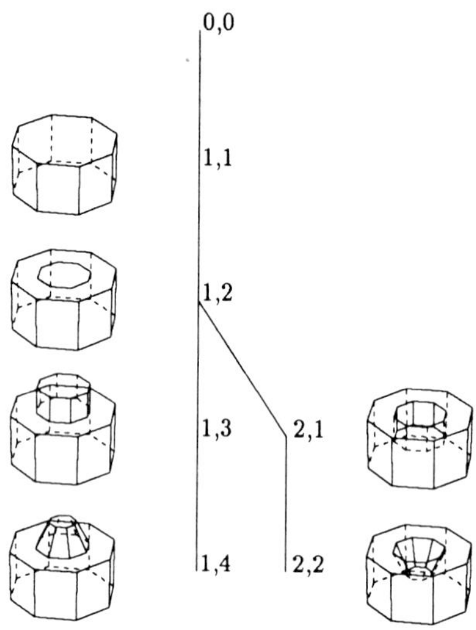

Fig. 10.2. Correspondence of design history tree and shape

As shown in Figure 10.2, a node in the design history tree corresponds to a shape of a solid at a particular stage; an arc that connects two nodes corresponds to the modification operation, that is, a command. A **state number** is automatically labeled to a node to indicate its corresponding shape. A state number consists of a pair of numbers: one called the **stage number** and the other called the **step number**. The stage number increases by increments of one as the stream of design changes and the step number increases similarly as a modification operation is performed in a design stage. This system identifies the order of the modification on all shapes in a design history tree. The design history tree in Figure 10.2 shows that the stream of design has changed because of the UNDO operation, and so the state number changes from 1,2 to 2,1. The design process should be represented by the tree structure since all the past processes exist in the design history.

In an arc of a design history tree, a sequence of primitive operations which modifies the solid shape, is stored. In DESIGNBASE, every primitive operation has its reverse operation. The reverse operation is also a primitive operation. If the reverse operations of a sequence stored in an arc are executed in the reverse order, then the operation corresponding to the arc can be canceled. This is how the UNDO operation is realized. On the contrary, the REDO operation is realized by executing the primitive operations of a sequence in the normal order, so that a shape lost by the UNDO operation can be retrieved.

The UNDO · REDO operations in DESIGNBASE can be regarded as those operations that change a shape to one that existed in the past, and not as the operations that cancel or redo modification operations. More concretely, the UNDO operation traces the design history tree from its roots and changes the solid shape into one that existed previously, and the REDO operation traces the design history tree towards its leaves and restores the shape lost by the UNDO operation.

Mäntylä explains the UNDO operation in Euler operations[90] and introduces a method called "inversion" as an application of the UNDO operation. This method stores the shape of a solid by storing the sequences of Euler operations that are executed to delete the solid. In order to reconstruct the solid shape, the UNDO operation of the stored sequences must be executed. With this method, the solid shape is stored as the procedure for reconstructing the shape, even though the shape itself is not stored.

10.3.3 Representation of the solid creation process

In DESIGNBASE, a design history tree is represented by using characters. The design history tree in Figure 10.2 is represented as shown in Figure 10.3.

```
0,0
 | gecy
1,1
 | prho
1,2 ______.
 | lifa   | lifa
1,3      2,1
 | scfa   | scfa
1,4      2,2 #
```

Fig. 10.3. Design history tree

The character strings on the right-hand side of the arcs in the design history tree are the names of the commands executed to modify the solid shape. The current state of the solid shape (in this case 2,2) is indicated by #. While checking the design process using the design history tree, the designer can proceed with the UNDO · REDO operations or by specifying the state numbers, he can recreate any shapes which previously existed.

Regarding labeling of the design process represented by the tree structure, Hoffmann uses a pair of numbers that correspond to a node[65]. One number in a pair indicates the depth of the tree structure and the other number indicates the order of creation of the nodes in that depth. This method can easily represent a tree structure but cannot represent the design order.

10.3.4 UNDO · REDO with Boolean operations

Hoffmann explains that the Boolean operations in a B-reps solid modeler are hardly suitable for the UNDO operation[65]. In DESIGNBASE, however, Boolean operations are suitable for the UNDO · REDO operations, because the system is so constructed that the Boolean operations are realized by using only the primitive operations[148]. For example, three kinds of Boolean operations are performed between a regular icosahedron (named I) and a regular dodecahedron (named D) as shown in Figure 10.4. The design history tree of these operations is created as shown in Figure 10.5. It is based on the solid I. The partner solid in a Boolean operation is also described in the design history tree on the right-hand side of the command name.

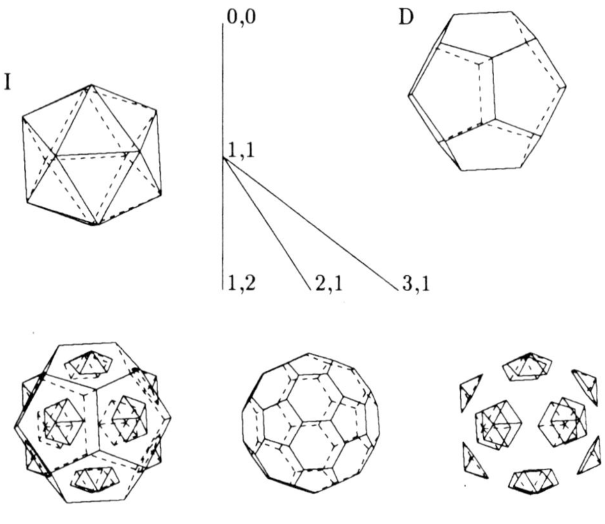

Fig. 10.4. Boolean operations

The procedure of the Boolean operations between solid I and solid D is indicated in the design history tree and is as follows: — first, the union operation is performed between the two solids and then the UNDO operation is performed. Next the intersection between the solids is obtained and the UNDO operation is again performed. Lastly, the difference between the solids is obtained. The shape given by the state number 2,1 is a quasi-regular polygon like a soccer ball.

After the difference is obtained in an operation like that given above, the shape may sometimes be topologically separate as shown by the state 3,1 in Figure 10.4. DESIGNBASE supports the operation to change the parts of a solid which are topologically separate into several independent solids. In DESIGNBASE, these operations, as well as the usual modification operations, are available in the UNDO · REDO operations.

10.4 Creation of Similar Shapes

10.4.1 Limits of the UNDO · REDO operations

DESIGNBASE supports the powerful function called UNDO · REDO to aid design where trial and error is needed. However, as explained earlier, the UNDO · REDO operations can only retrieve a shape that existed in a past stage of design.

```
0,0
 | geph
1,1 -----.-------.
unio D  inte D  diff D
1,2     2,1     3,1 #
```

Fig. 10.5. Design history tree representing Boolean operations

To overcome this limitation of the UNDO·REDO operations, DESIGNBASE also supports the reexecution function, which stores an executed command and its arguments in an arc of the design history tree. It is able to create a shape that did not exist in the past by executing the command with different arguments.

10.4.2 Reexecution function

It is evident that a shape corresponding to a node in the design history tree, can be created by executing all commands between the root and the node of the design history tree. If the numerical arguments of such a command are changed and the command is executed again, a shape different from the original shape is created.

Usually the topological elements of a solid should be referenced to execute a command. In the situation, where only the numerical arguments of a command are to be changed, the command can automatically be reexecuted by tracing the design history tree, without having to specify any other arguments. If an argument is to be changed, the state is traced back to the node that contains the argument by repeating the UNDO operation and then by reexecuting the commands stored in the arcs, then a different shape can automatically be created. As such, in DESIGNBASE, a shape similar to one that previously existed in the design process can be created by changing the numerical arguments of the commands in the design history tree.

Let us consider the example given in Figure 10.6. The solid on the left-hand side is at first created and then the rounding operation is performed to create the solid on the right-hand side.

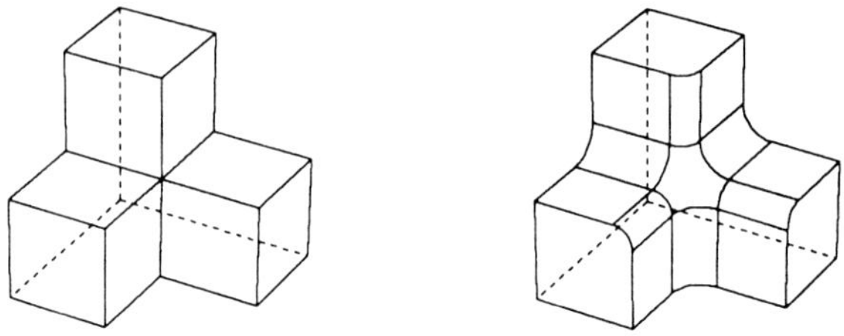

Fig. 10.6. Creation of a solid and its rounded edges

In doing this, six arguments are specified to determine the size of the solid and two arguments are specified for the rounding radii. The shape can be varied as shown in Figure 10.7 by changing the arguments for the rounding radii. If the arguments for the solid size are changed at the same time, the shape of the solid can further be varied as shown in Figure 10.8. Figure 10.9 shows the design history tree of these operations.

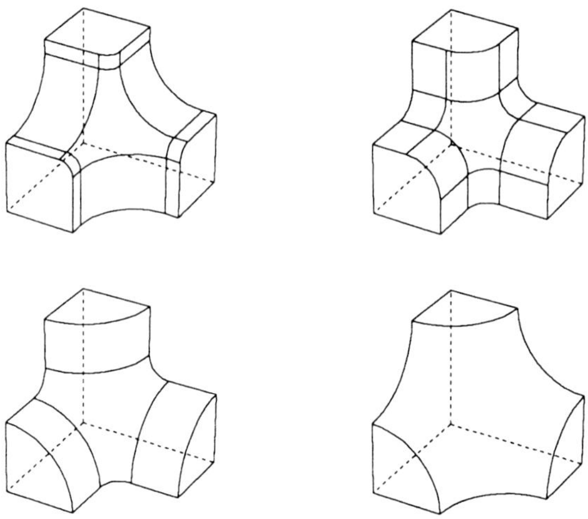

Fig. 10.7. Arranging of rounding radii (state 2,1 3,1 4,1 5,1)

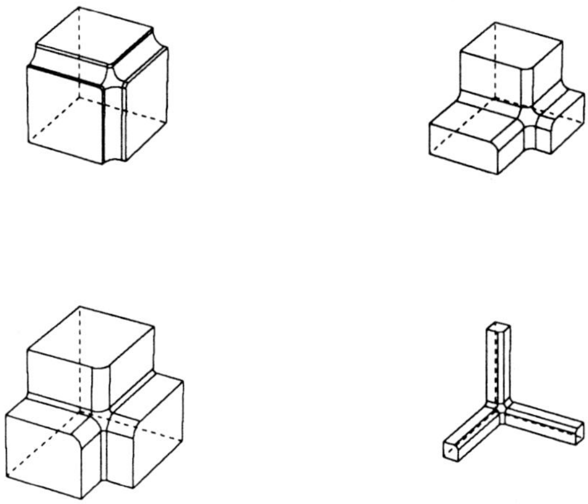

Fig. 10.8. Arranging of solid size and rounding radii (state 6,1 7,1 8,1 9,1)

```
0,0 _____._______._______._______.
 | gebl  | alte  | alte  | alte  | alte
1,1     6,1     7,1     8,1     9,1 #
 | lifa
1,2
 | lifa
1,3
 | lifa
1,4
 | deev
1,5 _____._______._______._______.
 | spre  | alte  | alte  | alte  | alte
1,6     2,1     3,1     4,1     5,1
 | spre
1,7
 | prcm
1,8
```

Fig. 10.9. Design history tree representing reexecution functions

The operation to modify solid shape by using the reexecution function is also stored in the design history tree. This means the UNDO·REDO operations can also be applied to such an operation. The design history tree in Figure 10.9 represents the state after the "alte" command, in which the reexecution function is implemented.

10.4.3 Limits of reexecution function

A shape similar to another shape can easily be created by effectively using the DESIGNBASE reexecution function. Usually, the reexecution is performed by merely using the data on the commands that were previously executed. Because of this construction system, if the topology of a solid is changed as a result of a change in the arguments of a command, the topological elements of the solid are also changed. Hence, the reexecution of the command may sometimes fail. In DESIGNBASE, however, although the command arguments might be changed, the UNDO·REDO operations are automatically performed so that a designer will not suffer from the trouble caused by change in the command arguments. This is one of the characteristics of the reexecution function in DESIGNBASE.

11. RENDERING

11.1 Introduction

When designing shapes in a 3-dimensional CAD system, the ability to easily evaluate the shapes during design is important to the designer. Some methods of doing this include: displaying contour lines, displaying control points, or eliminating hidden lines. The best method for showing the designed shape with realistic and persuasive images is **shading**. Since raster graphics technology was developed in the 1970's, shading was generalized, such that an image was divided into picture elements called **pixels** and displayed with the colors and brightness of each pixel. This method has enabled images to be displayed with reality on a screen.

The basic shading method is **constant shading**, which substitutes **polygons** (minute planes) for a solid. Each polygon is represented by one color in which a solid is represented by a set of polygons. However, the boundaries of the polygons are shaded unnaturally because of the different colors. To solve this problem, smooth shading methods were developed such as **Gouraud shading** which is based on the concept of color interpolation[60]. Plate 3 of the frontispiece is an example of Gouraud shading.

What is troublesome when creating a 2-dimensional planar image of a solid in 3-dimensional space is "hidden-surface elimination"[141]. The processing of hidden surfaces involves the elimination of the parts of a solid which are invisible when the solid is viewed from a certain viewpoint. In order to accomplish this, two methods were developed: the first is the **Z-buffer method** in which the depth (Z-value) from the viewpoint is compared between polygons and the second is the **scan line method** in which the data on an image is created for each scan line of a screen. These methods were efficient algorithms for eliminating hidden-surfaces.

In the 1980's, when more realistic images were required, the pixel was again the focus of attention as the picture element instead of the polygon. Whitted presented the **ray tracing method**[161], in which the color of a pixel is obtained by detecting objects which affect the pixel and the color. The development of the ray tracing method has, in general, enabled the representation of more realistic images, for example, in reflection, shadowing, or transmission refraction. In 1984, the **radiosity method** was introduced for the global lighting model. This method which applies the theory of heat transport in heat engineering[58] was proposed by Greenberg et al. The creation of a dim shade or light that can be seen in a room is one of the characteristics of this method. It has not been put into practical use because of the amount of calculation and memory capacity needed. In 1988, however, Cohen and Greenberg proposed an improved algorithm of the radiosity method[41], and it has become one of the noteworthy algorithms in the shading methods[72,78].

In addition to the various shading methods, there is a technique to attach raster data such as figures or textures to the surfaces of a solid to enhance the reality of an image. This technique is called **mapping** and has become indispensable in current computer graphics.

This chapter introduces shading models, hidden-surface elimination, and mapping, which are recognized as the basic techniques in computer graphics. A **shading model** is a model showing how the color of a point on a surface is calculated when there is a viewpoint and a light source. We will consider the Z-buffer method, the scan line method, and ray tracing method as ways of accomplishing hidden-surface elimination. Shading models are used to calculate colors on the visible surfaces obtained with these methods.

11.2 Color Components and Shading Model

11.2.1 Color representation

Human beings through their eyes can recognize the existence of objects and their shapes in the real world. This is because there are colors that distinguish one object from another. A person is able to sense colors because the light emitting from a light source hits the surface of an object and is reflected with intrinsic wavelength to the observer. To this extent, "color" and "light" can be treated in the same category for discussion.

To represent colors in a computer, the colors should be expressed by numerical values. The popular methods for this are the **RGB**(Red-Green-Blue)**system** and the **HCV**(Hue-Contrast-Value)**system**. The RGB system expresses the three primary colors of light (red, green, and blue) by numerical values of their respective brightness. The HCV system uses the three elements of light (hue, contrast, and value). For computer processing, the RGB system is most used because it is suitable for the display emitter. Actually, a full-color display represents a color by using 24 bits; 8 bits for each of R, G, and B. It can display about 16,700,000 (2^{24}) colors, a much larger number than human beings can distinguish themselves.

In addition to shapes and colors, one can recognize texture when viewing an object. A person can estimate what an object is made from by using several factors including the subtle shades on an uneven surface of an object, the highlights, i.e., where it is especially bright, or the transparency of the object. The difference in the appearance of an object, caused by the different materials (such as metal, glass, rubber, or paper), is the result of a physical phenomenon between the light and the surface of the object.

It is complicated to simulate all of these physical phenomena in a computer. Therefore, in computer graphics, these phenomena are abstracted using shading models expressed by numerical equations. The intensity of light, $\mathbf{I}$, that determines the color of the surface of an object is usually obtained by adding the **ambient light $\mathbf{I}_a$**, the **diffuse reflection light $\mathbf{I}_d$**, and the **specular reflection light $\mathbf{I}_s$**, as follows:

$$\mathbf{I} = \mathbf{I}_a + \mathbf{I}_d + \mathbf{I}_s \tag{11.1}$$

Each of these types of light is explained in the following subsections.

11.2.2 Ambient light

Imagine that there is a solid and a source of light. Besides a part of the solid which is directly lit by the source, you can also see another part of the solid in the shade, although it is not directly lit. This is because the part in the shape is lit to some extent by indirect light reflected from surrounding objects or walls and scattered light from the dust in the air. This light is called **ambient light**. Usually, to avoid complicated calculation, the value of the ambient light at any part on the surface of an object is assumed to be constant. In other words, the ambient light does not depend on the direction of a surface or the position of a person's eyes. It is expressed by the following equation:

$$\mathbf{I}_a = k_a \mathbf{I}_L \tag{11.2}$$

$\mathbf{I}_a$: Intensity of reflection of ambient light
k_a : Reflection coefficient of ambient light
$\mathbf{I}_L$: Intensity of light source

11.2.3 Diffuse reflection light

Some of the incident light on an object is reflected and diffused between particles of the object and a part of it leaves the object. This light is called the **diffuse reflection light**. The diffuse reflection light is uniform and independent of the direction of the viewpoint; thus its intensity is determined by direction of the light source. Figure 11.1 illustrates the relationship between a surface area S and the amount of incident light when light is directed toward the surface at the incident angle θ to the normal direction $\mathbf{N}$ of the surface. $\mathbf{L}$ is the vector that is directed toward the light source.

By using the Lambert model, which is the easiest shading model, diffuse reflection light $\mathbf{I}_d$ is expressed by the following equation:

$$\mathbf{I}_d = k_d \mathbf{I}_L \cos\theta \tag{11.3}$$

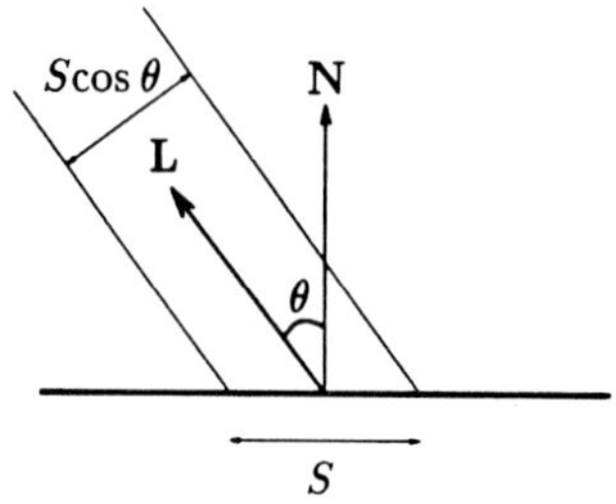

Fig. 11.1. Lambert model of diffuse reflection light

$\mathbf{I}_d$: Intensity of diffuse reflection light
k_d : Diffuse reflection coefficient
θ : Incident angle of the light

11.2.4 Specular reflection light

Light that is reflected from the surface of a metallic object includes **specular reflection light**. The intensity of this light can be expressed by using the specular reflection coefficient k_s. The texture of the surface of an object can be varied by changing the value of k_s, such as, the greater the value of k_s, the more the surface looks like well-polished metal.

We shall now consider the representative shading models of specular reflection light. These are the **Phong model**, **Blinn model**, and **Cook-Torrance model**.

Phong model

The **Phong model**[108] is the most popular shading model of specular reflection light. This light is usually reflected in the regular reflection direction **R** from the incident direction **L**; the model is obtained by supposing that the light is also reflected in the **E** direction (in proportion to $(\cos\theta)^n$), where θ differs from the regular reflection direction (Figure 11.2).

The model is expressed as follows:

$$\mathbf{I}_s = k_s\mathbf{I}_L(\cos\theta)^n \quad (11.4)$$

$$= k_s\mathbf{I}_L(\mathbf{E}\cdot\mathbf{R})^n \quad (11.5)$$

$\mathbf{I}_s$: Intensity of specular reflection light
k_s : Specular reflection coefficient
θ : Angle between the regular reflection direction to the incident direction and the line of sight
n : Scattering coefficient of specular reflection
$\mathbf{E}$: View vector

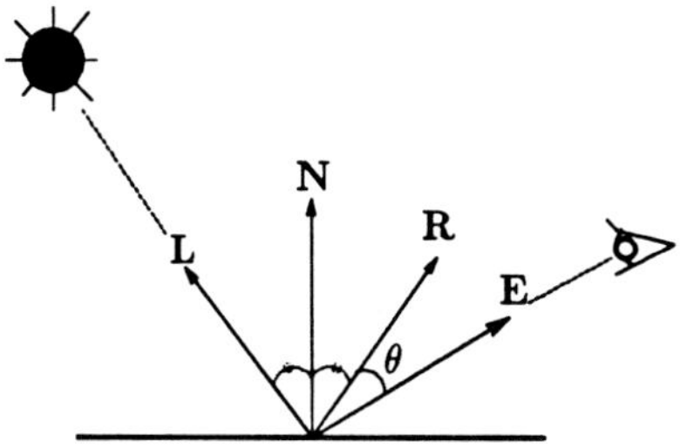

Fig. 11.2. The Phong model

The **specular reflection coefficient** k_s is actually the function $W(\theta)$ that varies with θ. The value of $W(\theta)$ is almost constant, when θ is changing between 70° and 80° only, so the constant k_s is usually used instead. As is obvious from Equation (11.5), when $\theta = 0$, that is, the viewpoint lies in the direction of regular refection, the intensity of the specular reflection light becomes maximum. On the other hand, when $\theta = 90°$, the intensity becomes 0. n is an amount unique to the object. The greater the value of n, the more the specular reflection light gathers in the regular reflection direction and the highlight effect is enhanced. This is one of the characteristics of the Phong model. The problem with this model, however, is that rough surfaces cannot be displayed well.

Blinn model

The **Blinn model**[14] is an improved Phong model because it deals with rough surfaces. This model is based on the supposition that the surface of an object is a group of minute faces of specular reflection light and the direction of the faces is random (Figure 11.3).

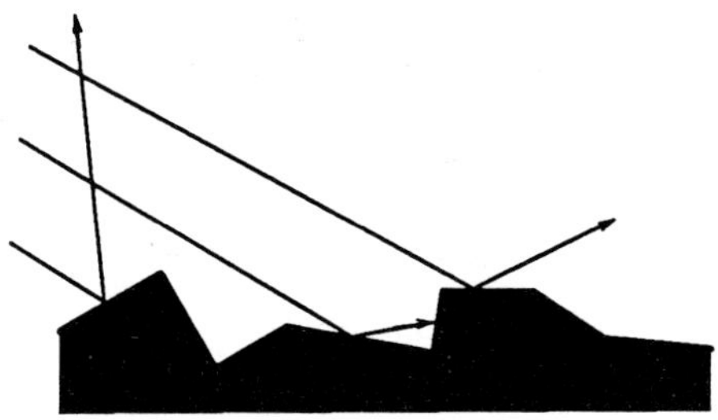

Fig. 11.3. A group of minute faces

This model is obtained by using the ratio of the minute faces that are in the **H** direction, where the angle between the incident light direction **L** and the view direction **E** is divided into two equal parts (Figure 11.4).

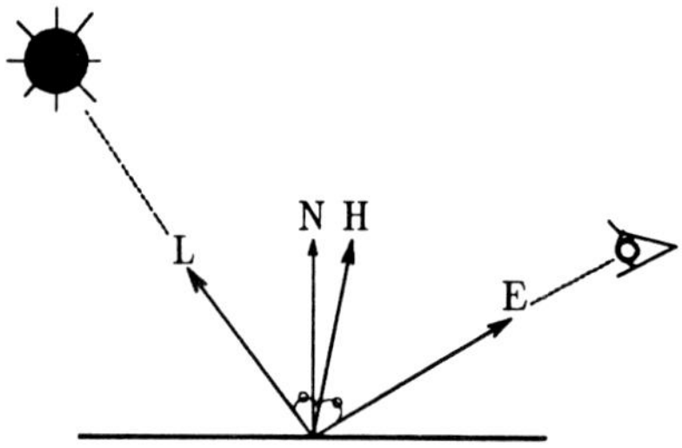

Fig. 11.4. Blinn model

Here, the direction **H** is defined as

$$\mathbf{H} = (\mathbf{E} + \mathbf{L})/|\mathbf{E} + \mathbf{L}| \tag{11.6}$$

which gives the bisector of the angle between the incident light direction **L** and the view direction **E**, where

$$|\mathbf{E}| = 1, \ |\mathbf{L}| = 1, \ |\mathbf{H}| = 1 \tag{11.7}$$

If the normal vector of a face is defined as **N**, then, the intensity of specular reflection light can be expressed by using the **distribution function of minute faces** D as follows:

$$\mathbf{I}_s = \frac{DGF}{|\mathbf{N} \cdot \mathbf{E}|} \mathbf{I}_L \tag{11.8}$$

$\mathbf{I}_s$: Intensity of specular reflection light
D : Distribution function of minute faces
G : Geometric attenuation coefficient
F : Reflection coefficient by Fresnel's formula

Fresnel's formula is set as follows:

$$F = \frac{1}{2} \frac{(a-b)^2}{(a+b)^2} \left\{ 1 + \frac{(b(a+b)-1)^2}{(b(a-b)+1)^2} \right\} \tag{11.9}$$

$$\begin{aligned} a &= \sqrt{n^2 + b^2 - 1} \\ b &= |\mathbf{V} \cdot \mathbf{H}| \end{aligned}$$

n : Refraction constant

G should be set to the smallest of the following three values:

$$\begin{aligned} G &= 2(\mathbf{N} \cdot \mathbf{H})(\mathbf{N} \cdot \mathbf{L})/(\mathbf{E} \cdot \mathbf{H}) \\ G &= 2(\mathbf{N} \cdot \mathbf{H})(\mathbf{N} \cdot \mathbf{E})/(\mathbf{E} \cdot \mathbf{H}) \\ G &= 1 \end{aligned} \tag{11.10}$$

When the angle defined by **N** and **H** is β, the distribution function of minute faces D should be set to one of the following expressions,

Cosine

$$D = (\cos\beta)^c \tag{11.11}$$

Gaussian distribution function

$$D = \exp[-(\beta c)^2] \tag{11.12}$$

Spheroidal harmonics of distribution function

$$D = \left[\frac{c^2}{\cos^2\beta(c^2-1)+1}\right]^2 c = |\mathbf{E}\cdot\mathbf{H}| \tag{11.13}$$

Cook-Torrance model

The **Cook-Torrance model**[43] is an improved Blinn model and is obtained by using the distribution function of minute faces D with the Beckmann distribution function as follows:

$$D(\beta) = \frac{1}{m^2\cos^4\beta}\exp\left[-\left\{\frac{\tan\beta}{m}\right\}^2\right] \tag{11.14}$$

where m is a constant which lies approximately between 0.2 and 0.7. The smaller the value of m, the narrower the range of the distribution. The minute face distribution can be obtained by adding the Ds obtained from the above equation with various ms. This model is more suitable for representing metallic texture than the Blinn model.

11.2.5 Transmission light

When shading a transparent or semitransparent object, **transmission light** which is the light that transmits through the object must be taken into consideration. The Newell model, a popular model of transmission light, is defined by adding the intensity of light reflected from a (semi)transparent object and the intensity of light transmitted through the object in a certain ratio. It is expressed as follows:

$$\mathbf{I}_t = (1-k_t)\mathbf{I}_f + k_t\mathbf{I}_b \tag{11.15}$$

$\mathbf{I}_t$: Intensity of transmission light
k_t : Transmission constant
$\mathbf{I}_f$: Light reflected from a (semi)transparent object
$\mathbf{I}_b$: Light transmitting the (semi)transparent object

This is a simple model, in which refraction is not considered. Should the refraction be considered, the ray tracing method would have to be used (see Section 11.5).

11.3 Z-Buffer Method

The representative methods for eliminating hidden-surfaces are the Z-buffer method and the scan line method. These two methods have the characteristic that a solid is shaded in units of minute planes called polygons into which the surface of the solid is divided.

The **Z-buffer method** determines the front-to-back relationship of polygons divided into units of pixels. It is implemented by a very simple algorithm but can perform fairly exact hidden-surface elimination[27]. The algorithm is as follows:

1. Allocate the memory storage for a buffer to store the depth data, i.e., distance from the viewpoint, that has one-to-one correspondence with a frame buffer. The **frame buffer** is a memory device in which addresses that correspond to the coordinates on the screen are stored. The depth is defined by the z-value in a coordinate system; thus the buffer that is prepared is called the **Z-buffer**.

2. Start all the frame buffers in the background color and begin with each value in the Z-buffers at infinity.

3. Perform the following operations for every polygon:

   (a) Set a Z-buffer value which corresponds to a pixel at (x, y) in a polygon to $buf(x, y)$ and the z-value for (x, y) in the polygon to $z(x, y)$.

   (b) If $z(x, y) < buf(x, y)$ is satisfied, transform the Z-buffer value into $z(x, y)$ and store the color of the polygon in the frame buffer (see Figure 11.5).

The algorithm of the Z-buffer method is quite simple, and therefore it can easily be incorporated into hardwares or framewares. On the other hand, the memory capacity for Z-buffers is relatively large. Moreover, one pixel has only one z-value and one color. Hence there is **jaggedness** at the boundaries of the polygons. Transparent objects also cannot be displayed by the Z-buffer method.

11.4 Scan Line Method

11.4.1 Characteristics of scan line method

The **scan line method** is an algorithm that eliminates the hidden-surfaces on a polygonized solid. This method creates image data in units of **scan lines**, hence the name is the scan line method. It saves on calculation, because a calculation in 3-dimensional space is at first converted into one on a 2-dimensional

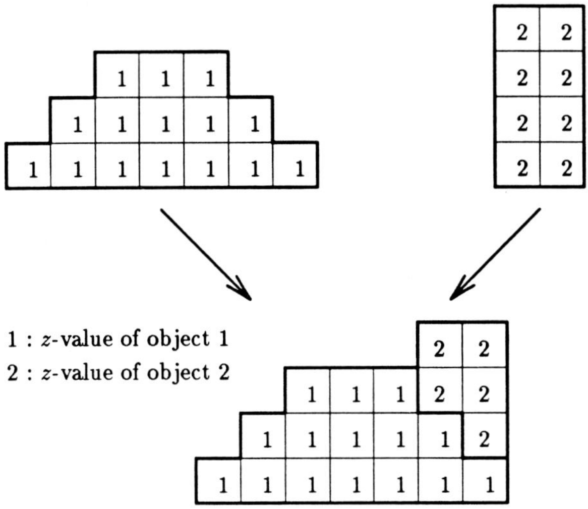

Fig. 11.5. Z-buffer method

plane (called **scan line plane**) which is defined by a scan line and a viewpoint. Then, the calculation is further simplified to one on the intersection curve (called a **segment**) between the scan line plane and a polygon.

The operation begins with the top scan line plane of the display. On a scan line plane, all polygons which intersect with the plane are selected. Then, in units of the **intersection segment** between these polygons and the scan line plane, the depth from the viewpoint to the polygons is compared. An enhancement of the scan line method is the **scan line–Z-buffer method**. In this method, depths are compared by using Z-buffers in units of segments in one line.

The scan line method can create a high quality image at relatively high speed. Furthermore, compared with the Z-buffer method, less memory capacity is required. Plate 4 in the frontispiece shows an image created by the scan line method in DESIGNBASE.

11.4.2 Actual processing in the scan line method

This subsection explains the concrete processing stream in the scan line method (Figure 11.6). The steps are as follows:

1. Polygonize a surface and transform the world coordinate system into a viewpoint coordinate system, and then into the screen coordinate system for easy calculation.

2. Obtain the maximum and minimum y-coordinates of every polygon whose normal vector has the angle with the view vector within 90°,

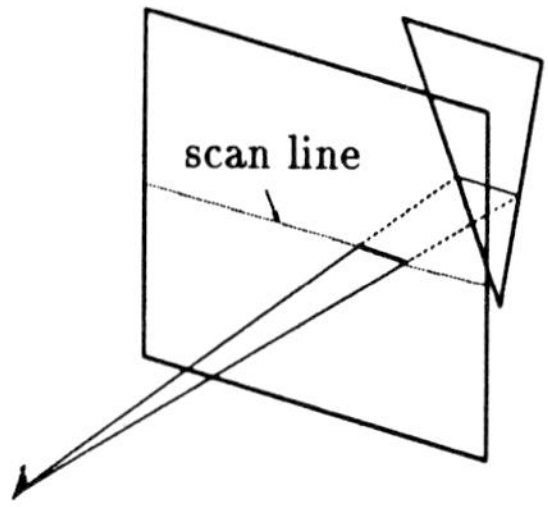

Fig. 11.6. Scan line method

then sort the minimum y-coordinates obtained in ascending order so as to sort the polygons.

3. Obtain the maximum and minimum y-coordinates of each polyline on the boundaries of all the polygons and then sort the minimum y-coordinates obtained in ascending order so as to sort the polygons (Y-sort).

4. Obtain the maximum and minimum x-coordinates of polylines that intersect with the scan line of the current y-coordinate, and then sort the minimum x-coordinates obtained in ascending order as as to sort the polylines in the active polyline list (X-sort). Then, from the active polyline list created, obtain segments, i.e. pairs of polyline data on both sides of polygons.

5. Sort the segment data in the active segment list in the ascending order about the z-coordinates (Z-sort).

6. Create the image data (color data) of one scan line.

7. To move the operation to the next scan line, update the active polyline list. If a y-coordinate of a scan line exceeds the maximum y-coordinate of a polyline, then the polyline is excluded from the data in the list. Repeat the operation from step 5.

11.4.3 Coherence in the scan line method

The scan line method can generate the relatively realistic image despite the small amount of calculation required, using the fact that the data on the x-coordinates, the normal vectors, and the view vector about a **polyline** on a given scan line is almost the same as that on the previous scan line. This is called **coherence** on the projection plane. When the data on the value of the first scan line and the difference in the data between scan lines are calculated, then it is not necessary to calculate the same elements for another scan line. Thus the amount of calculation required will be very small. This advantage makes it easy to update data in the active polyline list, during processing in the scan line method.

When the x-coordinate x_{i-1}, of a polyline on the $i-1$th scan line and the normal vector $\mathbf{N}_{i-1}$ of the polyline are known, the x-coordinate x_i on ith scan line and the normal vector $\mathbf{N}_i$ can be easily obtained from Equations 11.16 through 11.20 as such:

$$x_i = x_{i-1} + \Delta x \tag{11.16}$$

$$\mathbf{N}_i = \mathbf{N}_{i-1} \quad \text{(for plane)} \tag{11.17}$$

$$\mathbf{V}_i = \mathbf{V}_{i-1} + \Delta\mathbf{V} \tag{11.18}$$

$$\mathbf{L}_i = \begin{cases} \mathbf{L}_{i-1} + \Delta\mathbf{L} & \text{(for point light source)} \\ \mathbf{L}_{i-1} & \text{(for parallel light source)} \end{cases} \tag{11.19}$$

$$d_i = d_{i-1} + \Delta d \tag{11.20}$$

where x_i indicates coordinates, $\mathbf{N}_i$ indicates the normal vector, $\mathbf{V}_i$ indicates the view vector, $\mathbf{L}_i$ indicates the ray vector, and d_i indicates the distance from the light source. Here,

$$\begin{aligned} \Delta x &= (x_l - x_h)/(y_l - y_h) \\ \Delta\mathbf{V} &= (\mathbf{V}_l - \mathbf{V}_h)/(y_l - y_h) \\ \Delta\mathbf{L} &= (\mathbf{L}_l - \mathbf{L}_h)/(y_l - y_h) \quad \text{(for point light source)} \\ \Delta d &= (d_l - d_h)/(y_l - y_h) \end{aligned}$$

Figure 11.7 shows the polyline with both its end points (x_h, y_h) and (x_l, y_l). If the x-coordinate on $i-1$th scan line is obtained, then the coordinate on ith scan line can also be obtained by adding Δx to the original value.

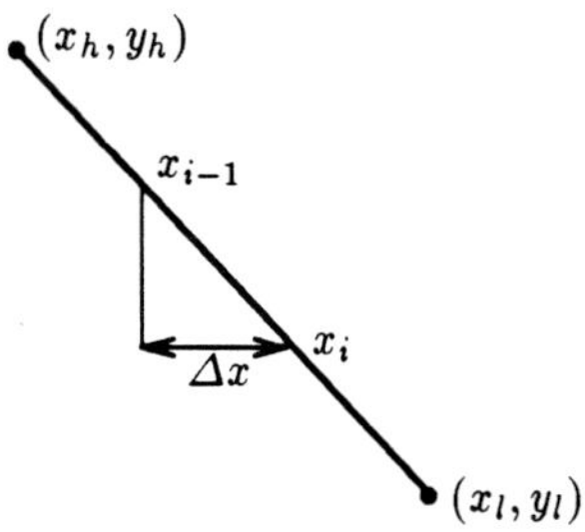

Fig. 11.7. Coherence in the scan line method

11.4.4 Antialiasing — A-buffer method

An image created with computer graphics may contain **jaggedness**, i.e. jagged boundaries. This phenomenon is called **aliasing** and the method to remove the jaggedness and to create a sharp image of high quality is called **antialiasing**. The **A-buffer method**[24] is one of the antialiasing methods of the scan line method, in which a pixel is divided into several subpixels and its color is determined by mixing the colors of the subpixels. The procedure of the A-buffer method is as follows:

1. Choose one pixel P (see Figure 11.8) that lies on the boundary of two polygons A and B.
2. Divide the pixel P into subpixels. In Figure 11.8, the pixel is divided into 32 (8×4) subpixels. The pixel in this condition is called the **pixel mask**.
3. Calculate the occupancy ratio S_A of the polygon A in the pixel P and S_B of the polygon B.
4. Calculate the following expression, in which the color of the polygon A is C_A and the color of the polygon B is C_B, so that the color of the pixel P can be obtained:

$$S_A \cdot C_A + S_B \cdot C_B \tag{11.21}$$

The A-buffer method can also apply to shading of transparent and semi-transparent objects.

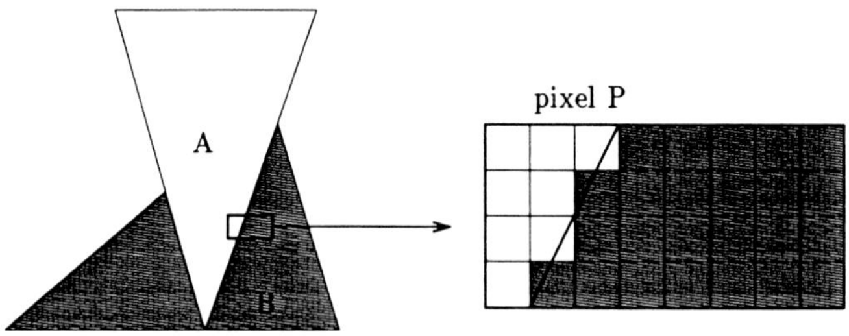

Fig. 11.8. A-buffer method

11.5 Ray Tracing Method

In the **ray tracing method**, a light ray is traced at first from the observer to each pixel on the screen. Then the intersection point between the light ray and an object is obtained, so that the color of the point is set as the color of the pixel. The color is further adjusted with reflections and refractions of the light ray. With this algorithm, high-quality images can be created with shadows, reflections, or (semi)transparency. Furthermore, every pixel is processed in the same way, which means that the same calculation is used every time in the ray tracing. Additionally memory capacity is saved. However, because calculation is done for each pixel, the amount of intersection calculation between light rays and objects will be large, and as a consequence this causes a long processing time in the ray tracing method.

Figure 11.9 illustrates how a light ray is reflected and refracted from an object and reaches the observer.

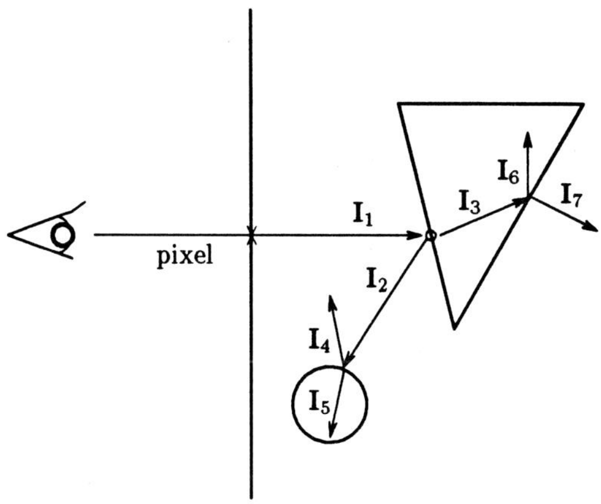

Fig. 11.9. Light ray travel to observer

The reflection and refraction of the light ray in Figure 11.9 is shown by the tree structure in Figure 11.10.

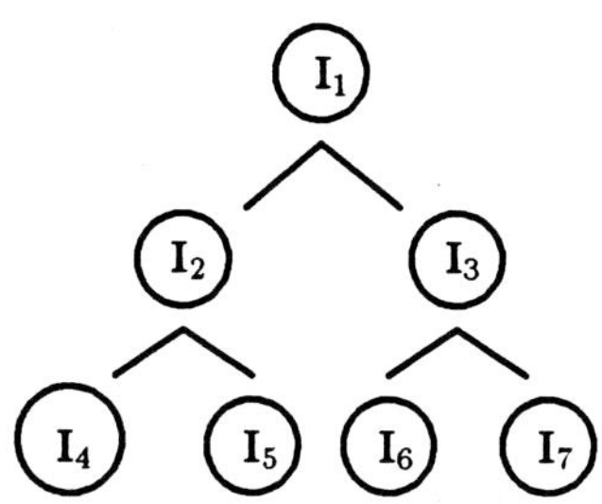

Fig. 11.10. Tree structure using the ray tracing method

The color of a pixel is determined by adding all of the elements of the pixel in a tree structure. When the reflected or refracted light ray no longer intersects with any object or when it reaches the light source, the tree terminates. In the actual processing, however, the calculation of reflection and refraction usually terminates when the depth of the tree structure reaches a level.

In the ray tracing method, the color $\mathbf{I}$ of a point on the object surface is obtained by adding the ambient light, the diffuse reflection light of each light source, the specular reflection light, and the transmission refraction light as follows:

$$\mathbf{I} = \mathbf{I}_a + k_d \sum_{i}^{l} |\mathbf{N} \cdot \mathbf{L}_i| + k_s \mathbf{S} + k_t \mathbf{T} \tag{11.22}$$

S : Intensity of reflection light
T : Intensity of refraction light

Plate 5 in the frontispiece shows a ray-traced image created by DESIGNBASE.

11.5.1 Antialiasing in the ray tracing method

Like the scan line method, the ray tracing method contains an antialiasing technique to enhance the quality of images. There are two antialiasing methods, which are set out as follows:

1. Produce several light rays about a pixel and average the result. The calculation time becomes longer in proportion to the number of light rays, but a fair image can be obtained.

2. Produce four light rays at the four corners about a pixel (see Figure 11.11(b)) and average the result. The light ray is usually produced from the center of each pixel as shown in Figure 11.11(a). With this antialiasing method, the light ray at one sampling point is used for its connecting four pixels, thus the density of sampling becomes quadrupled, while the actual number of light rays does not increase so much and the quality of the images is enhanced. This method can be said to be an efficient method for creating fair images with reasonable load for the computer.

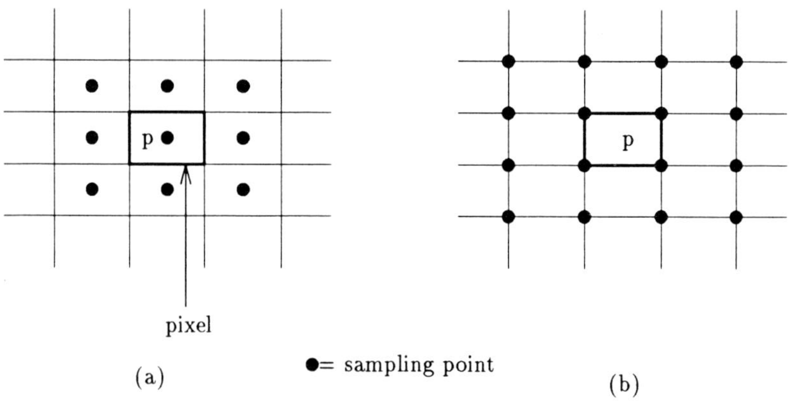

Fig. 11.11. Antialiasing in the ray tracing method

11.6 Mapping Method

11.6.1 Significance of mapping

As explained in Section 11.2, the difference in the texture of materials (such as metal, plastic, or rubber) can be expressed by changing reflection constants, various reflection coefficients, or shading models, in each shading algorithm. With these techniques, however, it is difficult to express, for example, a grain of wood, the pattern of marble, or the surface of sand. The **mapping** method was developed to realize such textures in computer graphics.

Mapping is that technique, very popular in computer graphics, for displaying various textures by attaching 2-dimensional raster data onto the surface of an object. Actually, when the color of a point on an object surface is calculated, the color is obtained from the raster data and is defined as that of the object. The following section introduces the three most popular mapping techniques — **texture mapping**, **reflection mapping**, and **refraction mapping** — and in addition, explains **bump mapping** and **solid texture**.

11.6.2 Texture mapping

Texture mapping is the technique used to express the texture of an object by attaching 2-dimensional raster data onto the surface of a solid. The 2-dimensional data used in texture mapping is called **texture** and it can be retrieved from the scanner or raster data created in advance. By mapping a logo or the texture of wood grain, for example, the surface of an object can be expressed with texture that could never be realized by the rendering technique only.

When performing texture mapping, a point $P(x, y, z)$ on the surface of an object in 3-dimensional space should correspond to a point $Q(s, t)$ in the texture coordinate system through a certain function F. Usually, a 2-dimensional coordinate system is defined in 3-dimensional space and the coordinate system is converted into the texture coordinate system. Unless the 2-dimensional coordinate system is properly defined, the texture mapped onto the surface of an object will be deformed. However, it is difficult to define the 2-dimensional coordinate system if the shape of the object surface. This problem can be solved by the **two-part texture** technique. In this technique, a virtual surface is first set onto the object surface. It is advisable to use a cylindrical or spherical surface, as the virtual surface, onto which the 2-dimensional coordinate system is easy to be defined. The texture is temporarily mapped onto the virtual surface, with the result that a point on the virtual surface is correspondent to a point in 3-dimensional space. Finally the color of the point can be determined.

The actual processing is carried out as follows:

1. Set a virtual surface of proper size such that it covers an object. It is helpful if various types of the virtual surface are prepared, e.g., cylindri-

cal surface, spherical surface, or planar surface. Use a cylindrical surface or spherical surface to cover the object and use a planar surface to lie on the object surface.

2. Map the mapping data onto the virtual surface. The data can easily be mapped because the virtual surface shape is simple. Note, that if a spherical surface is used, a distortion of data can be seen around the poles.
3. Extend a vector from a point on the object surface in the normal direction, so as to intersect with the virtual surface. The color of the virtual surface at the intersection point is defined as the color of the point on the object surface (see Figure 11.12).

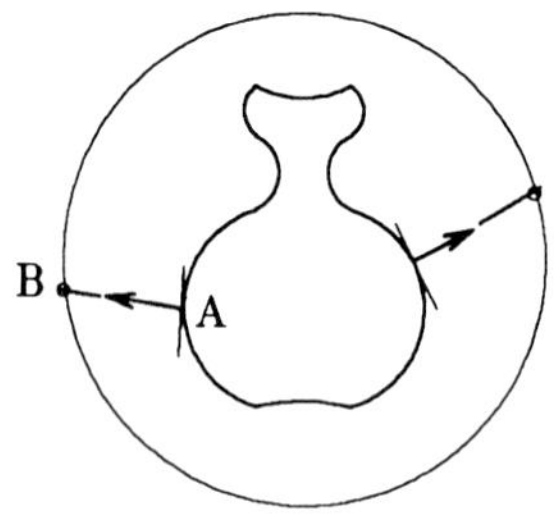

Fig. 11.12. Texture mapping (two-part texture mapping)

11.6.3 Reflection mapping

Surrounding objects are often reflected on the smooth surface of glass or metal. The reflection can be precisely calculated with the ray tracing technique, but it cannot be represented in the scan line method. To obtain a reflection effect in the scan line method, reflection data can be mapped onto the object surface. This technique is called **reflection mapping**. The processing is carried out as follows (see Figure 11.13):

1. As in the two-part texture mapping technique, set a virtual surface to cover an object or to lie on it.
2. Map the image data to the inside of the virtual surface.
3. Extend the view vector to the object surface, so that the vector reflects at the point A in the regular reflection direction, then reaches the point B on the virtual surface. The color of the point A is defined by blending the colors of these two points A and B.

As is clear from the above explanation, in the reflection mapping technique, only one reflection is considered, whereas in the ray tracing technique, it is more than one.

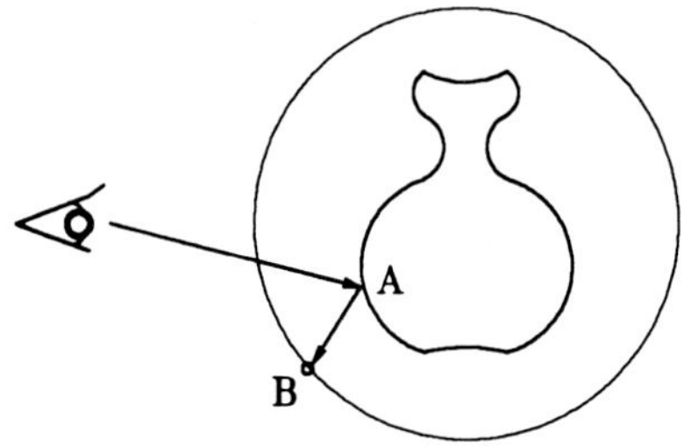

Fig. 11.13. Reflection mapping

11.6.4 Refraction mapping

Reflection of light rays is expressed by the reflection mapping technique. Similarly, refraction is expressed by the **refraction mapping** technique. With this mapping method, transparent or semitransparent objects can be expressed by mapping the data of transparent texture.

In the refraction mapping technique as well as the reflection mapping technique, concern is only with the image data on the virtual surface; hence the data is never influenced by surrounding objects. Refraction mapping is carried out as follows (see Figure 11.14):

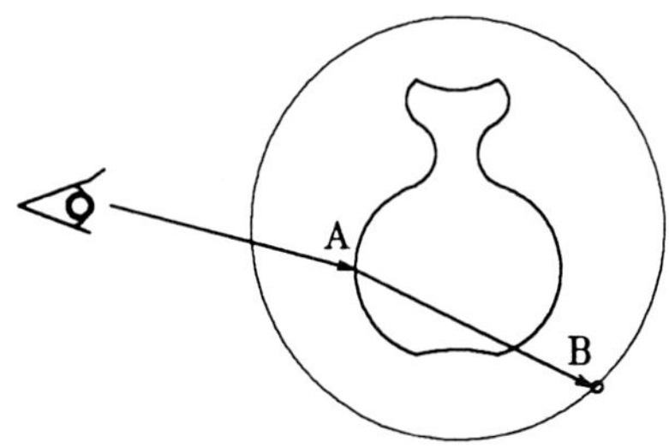

Fig. 11.14. Refraction mapping

1. Like in the two-part texture mapping, set a virtual surface to cover an object.

2. Map the image data to the inside of the virtual surface.

3. Extend the view vector to the object surface so that the vector refracts at the point A, then reaches the point B on the virtual surface. In the case of data on a transparent object, the color of the point A is defined by blending the colors of these two points A and B.

11.6.5 Bump mapping

Color data is mapped to the surface of an object as in the mapping techniques explained in Subsections 11.6.2 to 11.6.4. In bump mapping, instead of color, irregularity is mapped onto the surface of an object by using the difference of the normal vectors on the surface.

When an object is shaded by the scan line technique, the color of a point on the surface is calculated by using the view vector, the normal vector, and the ray vector at that point. Actually, the normal vector at a point in the mapping data is added to the original normal vector at the point and the color is defined. With this technique, bumpy or rough surface texture can be expressed with much reality. Note that the silhouette of an object does not have irregularity because the surface shape is not changed.

11.6.6 Solid texture

In Subsections 11.6.2 to 11.6.5, the surface of an object is the target of mapping. In the **solid texture** technique, the color data of a pattern created by some function is specified to some defined volume. Then the volume is sculptured, so that the contents of an object as well as its surface can be expressed with the specified patterns[104,106]. For example, a grain of wood or the pattern of marble can be expressed with the solid texture technique. Unlike the techniques in which 2-dimensional data is mapped to 3-dimensional shape, no distortion is caused, but patterns are difficult to modify because of the difficulty in controlling the functions.

12. MASS PROPERTIES

12.1 Introduction

One of the most important differences between solid models and surface or wireframe models is the stuffed contents. These give various properties to a solid model. This chapter explains how to calculate the basic mass properties of solids, namely the surface area, volume, center of gravity, and moment of inertia. As volume, center of gravity, and moment of inertia can all be obtained by using very similar methods, they are explained under one heading in this chapter.

12.2 How to Obtain Surface Area

In general, surface area is defined as follows:

$$A = \sum a(S)$$

This expression indicates the addition of the areas a of the surfaces S of a solid. It implies that the surface area calculation depends on the efficiency of obtaining the area of each surface. When obtaining the area of a surface S, the method will differ according to the type of surface; that is, for example, whether it is a plane or a curved surface. The popular calculation methods for planar surfaces are the **Monte Carlo method** and the **element subdivision method**. For the curved surface, the area of which is obtained through integration, this section explains **polygonization** and the **numerical integration method**[98].

12.2.1 Monte Carlo method

In the **Monte Carlo method**, a region S whose area is to be obtained is first enclosed with another region A, whose area is already known, as shown in Figure 12.1.

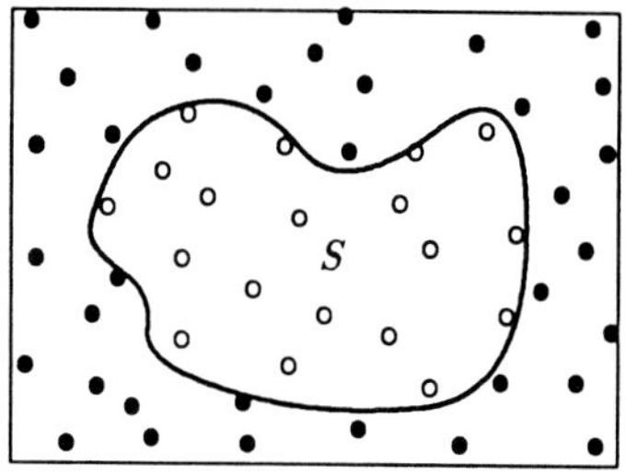

Fig. 12.1. Surface area calculation by Monte Carlo method

Now, suppose the region of the area A is marked at random with N dots and the number of dots h are lying in the region S. Then the area of the region S is obtained by

$$\frac{h}{N}A \tag{12.1}$$

Because the Monte Carlo method does not depend on the shape of a region whose area is to be obtained, it can be applied to various cases. The implementation of this method is also easy. However, high precision cannot be expected with this method; if precision is to be raised, the number of dots must be increased, and so the calculation time also increases.

12.2.2 Element subdivision method

The **element subdivision method** as the name implies, obtains the surface area, by subdividing a region S, whose area is to be obtained, into several parts for which the calculation can be easily done. For example, as shown in Figure 12.2, a region is divided into a polygon and a curved area.

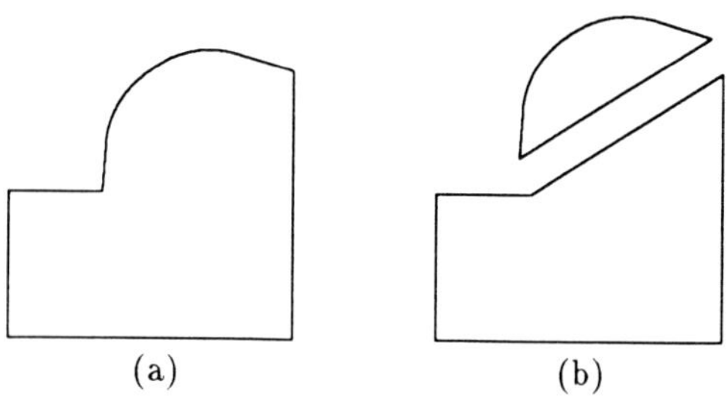

Fig. 12.2. Surface area calculation by element subdivision method

The area of the polygon can easily be obtained. If the coordinates of the polygon are defined to be

$$(x_1, y_1), (x_2, y_2), \cdots, (x_n, y_n)$$

then the area can be obtained as follows:

$$\begin{aligned} \text{area} &= \frac{1}{2}\{(x_1 + x_2)(y_1 - y_2) + (x_2 + x_3)(y_2 - y_3) + \cdots \\ &\quad + (x_n + x_1)(y_n - y_1)\} \end{aligned}$$

The calculation of the curved area is also easy by numerical integration if the expression for the curve is known. The Monte Carlo method is also able to obtain this area. By adding every area calculated, the total area can be obtained. This method can obtain the area of most regions precisely. Compared with the Monte Carlo method, the element subdivision method gives a high precision.

12.2.3 Polygonization

Polygonization is a method for obtaining the area of a surface by using polygons to approximate the surface. Figure 12.3 provides an illustration. In

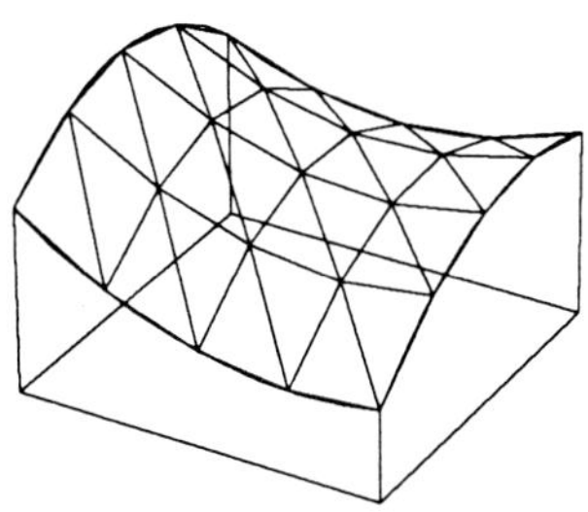

Fig. 12.3. Surface area calculation by polygonization

this example, the surface is polygonized with small triangles. If rectangles are used instead, the surface is integrated by using the trapezoidal rule.

In this method, the more the number of polygons used for approximating a surface, the higher the precision but the longer the calculation time becomes.

12.2.4 Numerical integration

Numerical integration is the method used to obtain the definite integral of the given function $f(x)$:

$$\int_b^a f(x)dx$$

Theoretically, except the the errors of numerical integration, this method can calculate the area of a curved surface precisely.

When a surface S is given by

$$z = f(x, y), \quad (x, y) \in D$$

the area $|S|$ can be obtained by the double integral in both x and y directions as follows:

$$|S| = \iint_D \sqrt{1 + \left(\frac{\partial z}{\partial x}\right)^2 + \left(\frac{\partial z}{\partial y}\right)^2}\, dxdy \tag{12.2}$$

Now, suppose the surface S is written by using the parameter (u, v) as

$$x = f(u, v), \quad y = g(u, v), \quad z = h(u, v), \quad (u, v) \in B$$

where f, g and h are continuously differentiable about u, v. Then, no singular point exists, and

$$\begin{aligned} J &= \begin{vmatrix} \frac{\partial x}{\partial u} & \frac{\partial x}{\partial v} \\ \frac{\partial y}{\partial u} & \frac{\partial y}{\partial v} \end{vmatrix} \\ &= \frac{\partial(x, y)}{\partial(u, v)} \end{aligned}$$

is proven. Under these conditions, the area $|S|$ of the surface S can be given as

$$|S| = \iint_B \sqrt{\left(\frac{\partial(y, z)}{\partial(u, v)}\right)^2 + \left(\frac{\partial(z, x)}{\partial(u, v)}\right)^2 + \left(\frac{\partial(x, y)}{\partial(u, v)}\right)^2}\, dudv \tag{12.3}$$

This area of a surface can be obtained by solving this equation.

Simpson's formula is one of the methods used to perform numerical integration in a computer. When the interval for integration $[a, b]$ is divided into m equal parts, Simpson's formula is given by

$$\begin{aligned} \int_a^b f(x)dx &\cong \frac{1}{3}hy_0 + \frac{4}{3}hy_1 + \frac{2}{3}hy_2 + \frac{4}{3}hy_3 + \frac{2}{3}hy_4 + \cdots \\ &\quad + \frac{2}{3}hy_{m-2} + \frac{4}{3}hy_{m-1} + \frac{1}{3}hy_m \\ &= \frac{h}{3}\left(y_0 + y_m + 2\sum_{i=1}^{m/2-1} y_{2i} + 4\sum_{i=1}^{m/2} y_{2i-1}\right) \end{aligned}$$

where

$$h = \frac{b - a}{m}$$

In addition to Simpson's formula, the trapezoidal rule and Gaussian formula are also available.

12.3 How to Obtain Volume, Center of Gravity, and Moment of Inertia

The representative methods for calculating volume, center of gravity, and moment of inertia are as follows:

- The Monte Carlo method
- The element subdivision method[102,84,85]
- The integration theorem

In general, volume, center of gravity, and moment of inertia are defined as

$$I = \int_V f(p)dV$$

This equation indicates the integration of a function f at all points p, where $p = (x, y, z)$ in a certain domain V. Here, the relationship between I and the function $f(p)$ can be defined as follows[84,85]:

$$\left.\begin{array}{ll} \text{If } f(p) = 1, & \text{then, } I \text{ indicates the volume.} \\ \text{If } f(p) = \dfrac{x}{m}, & \text{where } m \text{ is mass, then, } I \text{ indicates the } x\text{-coordinate of the center of gravity.} \\ \text{If } f(p) = x^2 + y^2, & \text{then, } I \text{ indicates the moment of inertia around the } z\text{-axis.} \end{array}\right\} \quad (12.4)$$

As is clear from these equations, if the procedure to obtain volume is provided, then the center of gravity and moment of inertia can be obtained by rearranging this procedure a little. Since these three properties can share most of procedures, it is advisable to calculate them all at the same time and save on the calculation time.

12.3.1 Monte Carlo method

Volume can be obtained by using the Monte Carlo method, which applies the method for obtaining surface area S, to a region V (refer to Subsection 12.2.1). Like in the surface area calculation, V is enclosed by another region whose volume is already known and the entire region is marked at random with many dots. While the surface area is obtained from Expression 12.1, volume, center of gravity, and moment of inertia are obtained from

$$I = \frac{h}{N} \sum_{i=1}^{N} f(p_i)$$

where the relationship between I and $f(p)$ follows Equation (12.4).

This calculation method has almost the same advantages and disadvantages as those in surface area calculation. In volume calculation, however, calculation costs must be further considered in determining whether points p lie inside or outside the region V. In a CSG-based representation, the determination is relatively easy; in a B-reps based system, the calculation cost for the determination becomes very high. For this reason, the Monte Carlo method is not very effective in a B-reps based system.

12.3.2 Element subdivision method

This method is accomplished by subdividing a solid and calculating the property for each subdivided part, then adding the results together.

For example, the solid S shown in Figure 12.4(a) is subdivided into parts b_1, b_2, b_3, and b_4, and its mass properties are calculated using the following equation:

$$\begin{aligned} I &= \int_S f dV \\ &= \sum_{i=1}^{4} \int_{b_i} f dV \end{aligned}$$

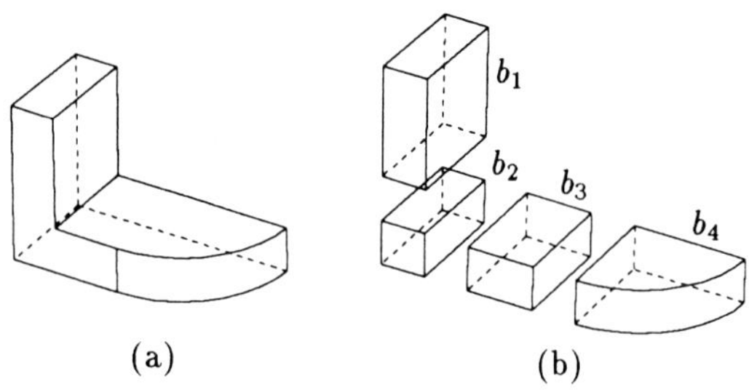

Fig. 12.4. Mass property calculation by element subdivision method

The efficiency of this method depends on how the solid is subdivided, although it is very difficult to subdivide a solid into several parts in order that the most efficient calculation is performed. Furthermore, If a part of the solid contains curved surfaces, its mass properties cannot be easily obtained.

The **octree structure** technique is used to compensate for this difficulty in subdividing a solid. It is one of the methods used to recursively subdivide a solid into eight equal cubic parts and then express the solid as a group of the subdivided cubes (see Figure 12.5). This technique is very systematic and is relatively easy, compared with the division shown in Figure 12.4. With it, the size of the error can be estimated. This estimation is based on one of the characteristics of the octree structure, that is, that the exact occupancy value of a cube which partly occupies a solid lies between the value for fully occupied cubes and empty cubes in the solid. Notwithstanding this, the octree structure has its problems: the numerical error is large because solid parts containing curved surfaces have also to be represented by cubes.

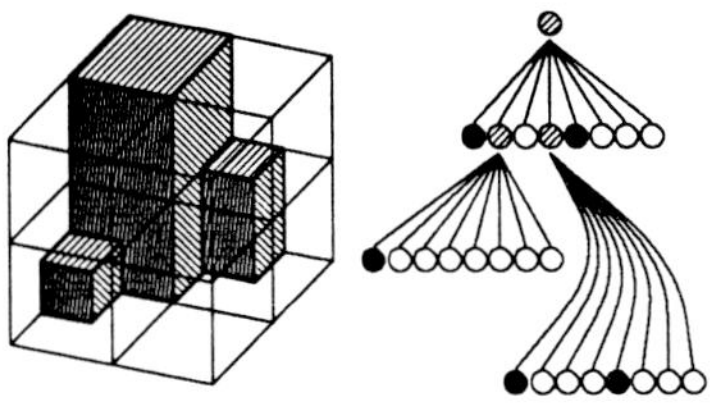

Fig. 12.5. Octree structure

12.3.3 Method using integration theorem

According to the Gaussian theorem, volume integral and surface integral have the following relationship about a region V with its surface area S:

$$\begin{aligned} I &= \int_V \nabla \cdot u \, dV \\ &= \int_S u \cdot n \, dS \end{aligned}$$

where n indicates the normal unit vector of the regional surface S and u indicates the vector function for determining the parameters for describing the mass distribution in the region V. By varying u, the following mass properties can be obtained.

$$\left.\begin{array}{ll} \text{If } u = \dfrac{\rho}{3}(xe_x + ye_y + ze_z), & \text{then, } I \text{ indicates the volume.} \\ \text{If } u = \dfrac{\rho}{2M}x^2 e_x, & \text{then, } I \text{ indicates the } x\text{-coordinate of the center of gravity.} \\ \text{If } u = \dfrac{\rho}{3}\left(y^3 e_y + z^3 e_z\right), & \text{then, } I \text{ indicates the moment of inertia around the } x\text{-axis.} \end{array}\right\} \quad (12.5)$$

In these equations, e_x, e_y, and e_z indicate unit vectors in each axis direction, M indicates the mass, and ρ indicates the density. By calculating the surface integral, the volume, center of gravity, and moment of inertia can be obtained. The surface integral can be calculated by using the numerical integral, as in the calculation of surface area (refer to Subsection 12.2.4).

One of the characteristics of this method is the ability to obtain mass properties from the surface S of a region V that represents a solid, that is, from the boundary data of the solid. Therefore, this method is more suited rather to B-reps based systems than to CSG-based system. Furthermore, although numerical error occurs because of the numerical integration on the surface, the error in calculation is basically small.

12.4 Mass Property Calculation in DESIGNBASE

Surface area, volume, center of gravity, and moment of inertia can all be calculated in DESIGNBASE. The methods selected for calculation are indicated in Table 12.1. These mass properties are obtained by integrating the data of curves and surfaces, which is easily acquired from boundary representations. Actual numerical calculation is done by using Simpson's formula of integration.

Table 12.1. Mass property calculation in DESIGNBASE

Surface area :plane	Element subdivision method (Numerical integration for curved parts)
:surface	Numerical integration
Volume	Integral theorem (Numerical integration)
Center of gravity	Integral theorem (Numerical integration)
Moment of inertia	Integral theorem (Numerical integration)

DESIGNBASE supports the trimmed surface as one means of representing curved surfaces. Since a trimmed surface represents only a part of its original surface, its shape is complicated, so that integration cannot be easily applied to it. Therefore, when a surface is a trimmed, Simpson's integral theorem is slightly modified and the numerical integration is performed as follows:

$$\int_a^b f_t(x)dx \cong \frac{h}{3}\left(T(y_0) + T(y_m) + 2\sum_{i=1}^{m/2-1} T(y_{2i}) + 4\sum_{i=1}^{m/2} T(y_{2i-1})\right)$$

$$T(n) = 0 \text{ or } n$$

where $T(y_i)$ is the function that returns y_i when the ith point lies inside of the trimmed surface or returns 0 when it lies outside of the trimmed surface. In the actual operation, when the integration of a trimmed surface is applied to the overall surface by using Simpson's integral theorem, the point lying outside the trimmed surface is ignored in the calculation. With this operation, psuedo-integration can be performed to a trimmed surface.

13. 3-DIMENSIONAL MODELING METHOD

13.1 Introduction

The solid modeler is one example in which structure analysis and NC processing have been applied in various forms. In recent years, attention has been focused on the resin model creation system, by which resin is cured in a 3-dimensional model shape. This chapter introduces the **resin model creation system** as another of the applications of the solid modeler.

13.2 Resin Model Creation System

The **resin model creation system** appeared in the market in the late 1980's. The system has been drawing attention since its introduction, because of its ability to supply 3-dimensional prototypes at lower costs in a shorter time, compared to conventional manual prototype or metal mold creation. This section explains at first what the resin model creation system is all about. There are other similar systems in the market and their fundamental principles are the same. The section then moves on to the basic technologies involved in the resin model creation system.

13.2.1 Fundamental principles

The resin model creation system creates resin models, using the **lamination method**. A resin model is formed by creating laminas from the **contour line data** and piling them up. This process is called **adhesion processing**, in contrast to the conventional stripping process, in which a 3-dimensional object is processed from its outside only, and its internal structure is not expressed. In adhesion processing, the internal structure of the object can be correctly expressed.

For the adhesion processing, the resin model creation system uses **ultra-violet-ray (UV) curing resin** as its material. The resin is irradiated with the UV laser, until it is cured, at which time the elevator moves down (see

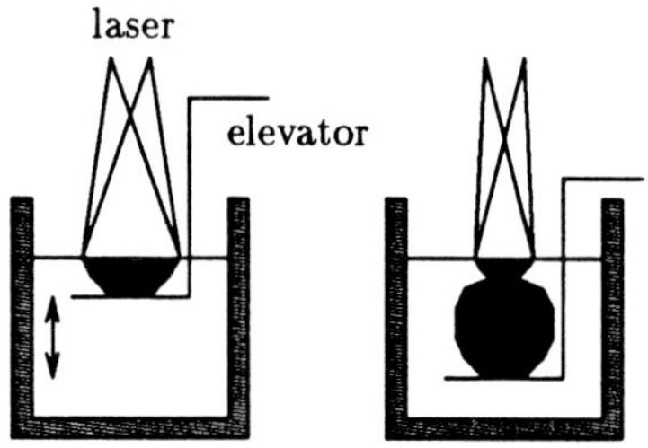

Fig. 13.1. Fundamental principle of resin model creation system

Figure 13.1). By repeating this operation, resin laminas are piled up and a solid model is created.

To accomplish this method, the following closely related technologies are necessary:

- UV curing resin
- Laser control
- Data processing

The UV curing resin should have many of the characteristics necessary for materials to create models. Laser control presents a range of problems, from the choice of laser ray to how to irradiate the resin. Data processing consists of two types of processing — one is the processing to convert solid data into a readable format for the resin model creation system and the other is the processing to delimit the area of laser irradiation and control the laser ray.

13.2.2 UV curing resin

At present, the following can be the constituents of the UV curing resin: urethane acrylate, epoxy acrylate, epoxy, and polyester acrylate. Even though the different resins should have the same basic characteristics, the one chosen depends on the system being employed. The characteristics are as follows:

1. Low shrinkage
   As resin is cured, the more it shrinks and the more the shape is strained; this leads to low precision. To avoid this problem, resin is required to have as little shrinkage as possible. Actually in the curing reaction, the resin which is irradiated is covered with liquid resin. Because of this technique, the low shrinkage is not so much required in resins (since there is sufficient resin to compensate for shrinkage). However, the lower the shrinkage, the more ideal the situation.

2. Low viscosity
   The viscosity of a resin affects precision and processing time. As one layer is cured, it becomes covered with resin which in turn is to be cured. If the viscosity of the resin is high, it takes a long time for it to become flat. Moreover, the center of the cured resin and its surroundings are swollen although it may look flat. If the swelling accumulates, the shape will be strained. On the contrary, if the viscosity is low, it takes a short time for the resin to be flat and as such precision is raised.

3. High curing ratio
   This is the degree to which the resin will be cured when irradiating ultraviolet-ray. If the ratio is low, then after the resin is cured and a lamina is created, the resin should be irradiated again by mercury lamp for further curing. The curing ratio of epoxy resin is high and no re-irradiation is necessary. Some resins are also cured without re-irradiation just so as to perform a rough check of shapes.

In addition to these characteristics, shock resistance after curing is also required. As the application range of resin models is spreading, the various characteristic requirements of the resins will be increased.

13.2.3 Laser control

The UV curing resin, as its name implies, is cured with ultraviolet-rays. As such, it may be cured by mercury lamp. However, high precision cannot be expected if a mercury lamp is used. Considering the process in which a laser is condensed and focused, it is better, for precision, to use the laser.

Lasers in the UV range are the helium-cadmium laser (He-Cd Laser) and the argon laser (Ar Laser). These lasers are selected for use depending on the system being employed. The output of He-Cd Laser is weak but it is easily handled and is fairly safe. On the other hand, the output of the Ar laser is strong but the handling is troublesome.

While various methods are suggested for scanning a laser ray in a resin, the two methods presently used are the **XY-plotter method** and **galvanometer-mirror method**. In the XY-plotter method, as its name implies, the laser ray scans the liquid surface in the same way as an XY-plotting. The galvanometer-mirror method reflects the laser ray with a mirror and the reflected ray scans the resin, while being controlled by the reflection angle of the mirror. In the XY-plotter method, the laser always scans perpendicularly to the liquid surface, while in the galvanometer-mirror method the laser scans the resin perpendicularly, directly below the mirror but it will scan slightly obliquely in another parts. This is the major difference between the two methods, which are compared in Table 13.1.

Table 13.1. Comparison of scanning method in XY-plotter and galvanometer-mirror

	XY-plotter method	Galvanometer-mirror method
Method	Vector/Raster	Vector/Raster
Speed	0.8 m/s	3.0 m/s
Range	Arbitrary, in theory	Small, in theory (about 300×300 mm)
Precision	High at arbitrary point	High at around center; otherwise, low

13.2.4 Data processing

As previously stated, data processing in the resin model creation system consists of two parts: processing to create contour line data according to the data created by a solid or surface modeler and processing to control a laser by delimiting the range of irradiation based on the contour line data created.

Of these two types of data processing, the processing to create contour line data is more difficult. The contour line is, as it can be seen in maps, the line which traces boundaries of the same height and always forms a closed curve. The major problems with creating contour lines are as follows:

1. A contour line is not closed.
   As shown in bold in Figure 13.2, when contour lines are obtained, they may sometimes have a jutting curve. This might occur because the data sent from a modeler is abnormal. The jutting curve should be deleted or should be copied to another contour line. This problem is discussed further in the following section.

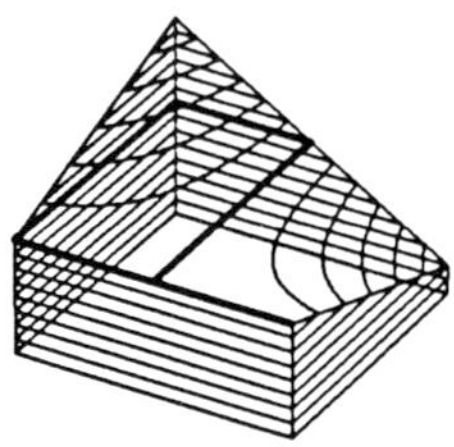

Fig. 13.2. Open contour line

2. A contour line is not uniquely defined.
   This problem relates to precision in processing. In Figure 13.3, for ex-

ample, the contour line of height H can either be that illustrated in (a) or (b), and the shape differs by a layer, say for example, 0.1mm.

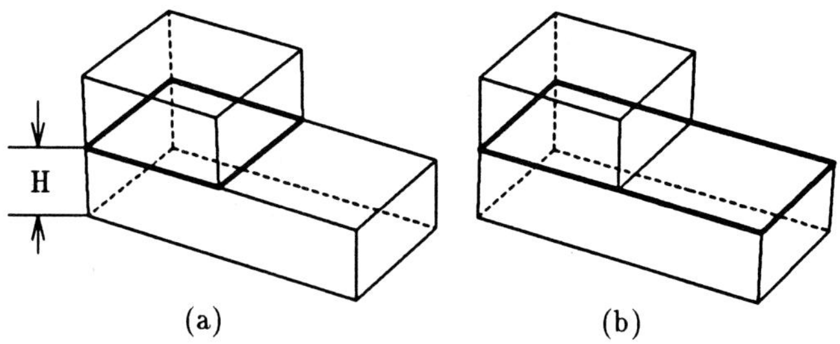

Fig. 13.3. Contour line which cannot be uniquely defined

3. The shape of a surface is complicated.
   Contour lines of a surface are intersection curves between the surface and a plane. As shown in Figure 13.4, when the shape of a surface is complicated, the intersection calculation itself is difficult. In this case, the calculation should be left to modelers.

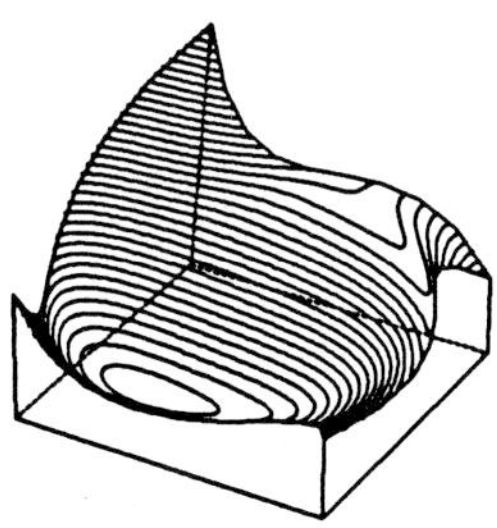

Fig. 13.4. Contour lines of a surface with complicated shape

These are the major problems which may occur when creating contour lines. However, these cases are special and thus, they are not found to be serious problems in an actual operation. In particular, it does not matter when the thickness of one-layer is almost the same as the processing precision in a resin model creation system, as shown in Figure 13.3.

After checking for these problems and generating the contour line data, the irradiation range of the laser is calculated according to the data, as shown in Figure 13.5. The range is calculated by the result of how many times a scan line intersects a contour line. Then the mirror or the coordinates for the XY-plotting are controlled. Theoretically, the data created by a surface modeler is able to create a solid resin model because the range is defined by using the surface data on the solid.

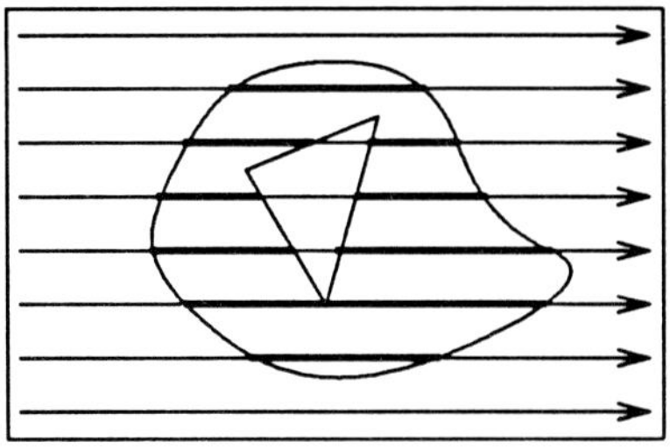

Fig. 13.5. How to define irradiation range of a laser

13.3 Link with Modeler

The resin model creation system does not incorporate the modeling functions in itself and is usually, linked to another modeler which sends data to it. The data received from modeler to create resin models will depend on the system being used and can be classified as follows:

- Triangular polygons and their normal vectors
- Surface data only
- Surface data and planar polygons
- Contour line data

Of these, the first data group (triangular polygons and their normal vectors) may contain errors more frequently than the other three groups of data.

With surface data, if the surface representation of a modeler differs from that of a resin model creation system, then errors may occur when approximating the data. Otherwise, the data can be sent to the system accurately and the processing precision can be controlled by the system.

When contour line data is sent, the data is accurate within a certain tolerance. In general, the system cannot raise the processing precision although it can lower it. Moreover, a precise resin model can be created if data is created by a solid modeler. Otherwise, the data, (especially data created by a surface modeler) may not sometimes be processed correctly. This happens because the data may not be correct; e.g., surfaces may be overlapping or intersecting. This is the reason why the data created by a surface modeler might not be processed in the resin model creation system. On the contrary, the data created by a solid modeler is rigidly precise, and contradictions in a solid shape, such as overlapping or intersection, never occurs. In conclusion, it is advisable to create the data by a solid modeler for sending it to the resin model creation system

Now we shall consider what kind of problems are raised when the resin model creation system is linked with a surface modeler or a solid modeler.

13.3.1 Link with surface modeler

The correctness of a solid shape should be the major consideration when linking the resin model creation system with a surface modeler. A surface modeler supports edge data in addition to surface data, and to this extent, a model is correct when processed in the surface modeler. However, surface data alone is not always sufficient to prevent contradictions in a model. For example, the following two problems are popular contradictions which prevent correct contour line data being created and thus defining the irradiation range (refer to Subsection 13.2.4):

1. Surfaces are not connected correctly.
   There are two cases of this situation: — one is when some part of a surface is jutting out as shown in Figure 13.6 (a), and the other is when two edges are not touching exactly as shown in (c). The contour line data of (a) which is shown in (b) contains an unnecessary line, and that of (c), shown in (d), is unclosed. Therefore, the correct contour line data cannot be created.

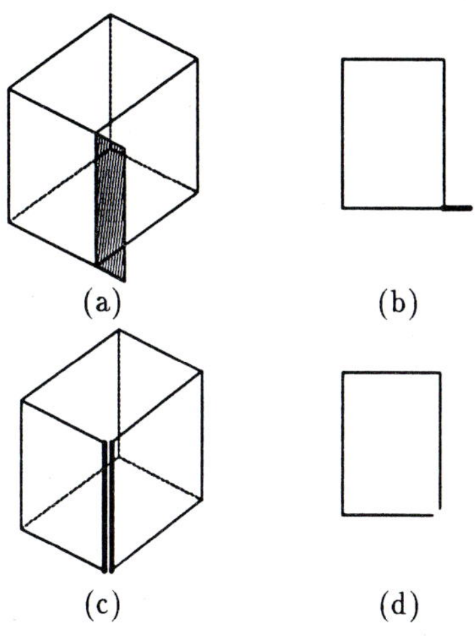

Fig. 13.6. Contour line data when surfaces are not connected correctly

2. Surfaces are overlapping.
   Two surfaces are overlapping, as shown in Figure 13.7 (a). The contour line is correctly closed in appearance, but the contour line data is duplicated where the surfaces are overlapping (Figure 13.7 (b)); hence correct contour line data cannot be created.

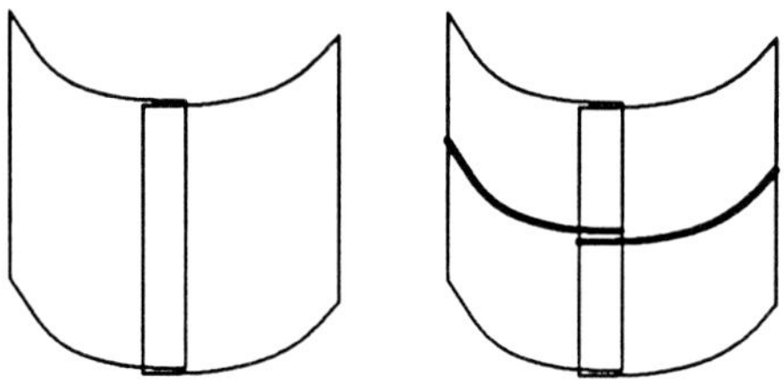

(a) (b)

Fig. 13.7. Contour line data when surfaces are overlapping

These problems can be avoided if a shape is carefully designed in a surface modeler. Nevertheless, they are substantial characteristics of surface modelers.

13.3.2 Link with solid modeler

Basically, there are no problems in linking the resin model creation system with a solid modeler, though the situation with trimmed surface must be pointed out. In recent years, trimmed surfaces have become popular as a means of surface representation in many solid modelers. The data on trimmed surfaces cannot be processed as it is, however, and so must be transformed in some other way. The following are the possible methods for transforming the data:

1. A trimmed surface is divided into repatched surfaces.
   A trimmed surface is divided into several repatched surfaces (refer to Section 8.2), as shown in Figure 13.8, and the data on the repatched surfaces is then sent to the resin model creation system. Generally, however, the division is difficult and sufficient precision cannot be expected.

2. Contour line data is created by a solid modeler.
   Because the resin model creation system cannot directly process a trimmed surface, contour line data is required to be created in the solid modeler to maintain the precision.

These problems can also be seen in the surface modeler.

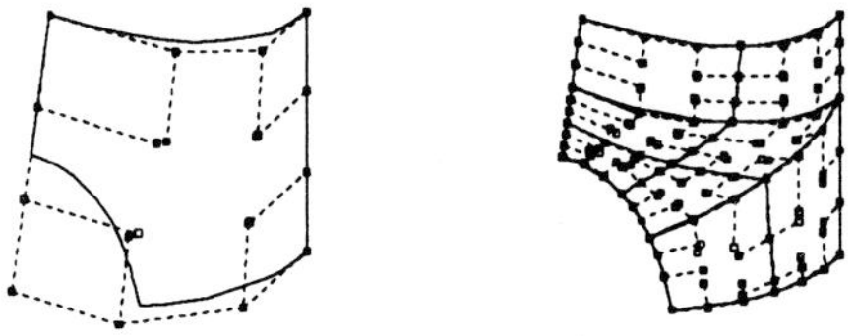

(a) (b)

Fig. 13.8. Division of a trimmed surface into repatched surfaces

13.4 DESIGNBASE and Resin Model Creation System

DESIGNBASE creates contour line data on a solid and sends it to a resin model creation system. The data can be retrieved from the system as it is. The following are the advantages of creating contour line data in the solid modeler:

- No error occurs due to surface conversion.
- Precision of contour line data can arbitrarily be modified.
- No dubious contour line data (unclosed or overlapping) is created.

In DESIGNBASE, surfaces may be trimmed surfaces, which can be processed in DESIGNBASE. Plate 6 in the frontispiece is an example of solid resin models whose data is created by DESIGNBASE.

14. SOLID MODELS AND STRUCTURE ANALYSIS

14.1 Introduction

Traditionally, prototypes of machine parts and industrial products were created by relying on a designer's perception and experience. The strength of the parts and products were also analyzed through manual experiments. In recent years, finite-element analysis has been improved from research to the practical use phase and is becoming an indispensable tool in research, design, and the development of engineering. Additionally with the rapid development in computers and their processing speed, finite-element analysis is playing an important role in shortening the period of product development, reducing the costs related to such development, and improving the reliability of products. This is not all, finite-element analysis can measure what could not be measured in experiments so far, according to the experimental conditions. This improvement has possibly created a whole new approach to design. Furthermore, by using super computers, analyses that had required long calculation time, such as large-scale problems or nonlinear issues, can be solved in a relatively short time. With the assistance of engineering work stations (EWS), soon finite-element analysis is also going to be realized in the personal environment.

This chapter addresses the finite-element method, the most popular in structure analysis, and its application of the finite-element analysis in solid modelers. KSWAD of Kubota Corporation is also explained as an example of an analysis system linked with DESIGNBASE.

14.2 History of Finite Element Method

The name **finite element method** was first used in 1960 by Clough in his paper on planar elastic problems[39]. However, the concept behind the method was not new, and different groups of applied mathematicians, physicists, and engineers have independently suggested the concept until present. This section describes the method with technological history.

Mechanics that originated with Newton, was developed by Euler and Lagrange, to give the method that analytically deals with phenomena related to energy. Based on this method, the theory of solid analysis was established. With this theory, the differential equation that represents a whole solid and its environmental condition could be obtained. By solving the equation, theoretically it became possible to analyze structures. With actual structures, however, the differential equation was found to be very complicated, making it difficult to be applied to the energy method. To compensate for this, methods to obtain approximate solutions by directly using the energy method were studied. One of these methods is the finite element method.

A structure that consists of straight-line members and has rigidity only in the axis direction is called a **truss**. It had already been known by the 1930's how to analyze trusses, that is, by combining each characteristic of its members according to the law of balance and solving the equations for the obtained load and displacement of the whole system. This method, however, could not be applied to a continuum such as a plate because a continuum is regarded as a structure that has an infinite number of nodes. Then, Hrenikoff[70] assumed that a continuous structure could be analyzed by conventional methods if it was divided into elements which were internally combined with only a finite number of nodes. The application of this assumption to actual analysis yielded success and saw the start of the finite element method in the engineering field.

Subsequently, Argyris et al.[1] improved the format of Hrenikoff's method so that it was easier to be processed in a computer and presented the method as the **matrix structure analysis method** in 1950. In 1956, Turner, Clough, et al.[150] of Boeing Company presented the application of the finite element method to box spar wings. This was the first paper on the finite element method in engineering. After Clough presented the paper on stress analysis of dams, the finite element method became popular and was developed for the aircraft industry in 1960's. Owing to the later development of the computer and the progress of its studies, the finite element method could not have been applied to this wider area. Reliability of the method has increased with experience, and at present the finite element method is the most powerful method in engineering analysis[157].

As super computers are developed, complicated analyses that could not have been accomplished earlier because of the high calculation costs, are being realized. Analysis techniques are also expected to be more precise with the expansion of super computers.

To get rid of low cost performance in the finite element method, a study on the **boundary element method**, introduced by the integral equation, is presently being conducted. Unlike the ordinary finite element method, this method does not divide the contents of an object into elements, hence input burden is greatly reduced.

14.3 Finite Element Method and Solid Models

This section briefly explains what the finite element method is about. Stress analysis of 3-dimensional structures is used as an example. Generally, the analysis of a structure involves a group of partial differential equations, which the calculation cannot be accomplished unless the structure is very simple.

At first, a structure is divided into small parts (finite elements). It is assumed that the internal displacement of these parts can be represented by simple polynomials of parameters that indicate their positions. The degree of the polynomials should be lowered to make calculation easier. A condition is set so that the stress between the adjacent elements is applied only to the points called **nodes**. Then, in order to obtain the coefficients of the polynomials, a simultaneous equation is established to express the relationship between adjacent parts; in this case, the principle of virtual work[157] is used. From the simultaneous equation the coefficients of the polynomials are obtained so that the structure can be analyzed[146].

Similarly, with the finite element method, a model must be divided into elements. This process is called **mesh generation** (**mesh division**). However, as finite element analysis is being put into practical use, the question has been raised as to whether complicated shapes should not be analyzed by modeling them instead of using divided elements. The mandays required for carrying out finite element analysis has also to be reduced, since 70 – 80% of the time is consumed in the preprocessing work such as mesh generation or condition setting. In an effect to achieve this goal, some systems have been able to assist in dividing structures with complicated shapes; they are called **preprocessors**. In recent years, more solid modelers are being used to input solid shapes into preprocessors.

When dividing a solid model into meshes, mesh size, that is, the fineness of the division and simplicity in shape should be considered. Since the finite element method uses approximations, parts of a model that have steep changes should be divided into smaller meshes in order to obtain the necessary precision. Note that too fine a division might increase discretization errors. On the other hand, parts which have moderate changes should be roughly divided; otherwise, the amount of calculation becomes very large and wastes calculation time and memory. The model to be analyzed should also be simplified. This is because solid models, often used in CAD systems, are also used as models for structure analysis. This means that data which might not be necessary for structure analysis will also be retrieved as the data on a model. Thus, the model should be simplified, with just enough data for the structure analysis[68].

The software that performs the analysis calculations is called the **solver**. The solver only calculates and obtains results; thus it is difficult to interpret the results from the numerical data obtained. The **postprocessor** was developed to address this problem. It transforms the numerical data into figures for easier understanding. Most of the present tools for structure analysis consist of the following two modules: a solver and **pre-postprocessor** which incorporates functions of both preprocessor and postprocessor.

14.4 3-Dimensional Mesh Generation

Many methods of generating meshes have been suggested, upon till now. The method to automatically decompose an object into meshes by using the given shape data is called **automatic mesh generation**. **3-dimensional automatic mesh generation** automatically creats a finite element model by specifying the mesh sizes of a solid model. Among the several methods of mesh generation, those that automatically generate meshes from solid models can be classified into the following three algorithms[64,123,132]:

- **Element extraction algorithm**
  This algorithm divides a solid model into meshes by using the topological data on the solid model. As shown in Figures 14.1, 14.2, and 14.3, tetrahedra are extracted from a solid model with three kinds of operations. The sizes of the meshes are not considered until the solid model is completely divided into tetrahedral elements. When the division is completed, each tetrahedron is decomposed into smaller tetrahedral finite elements.

- **Domain triangulation algorithm**
  This algorithm creates tetrahedral elements in a solid model, by distributing nodes within and on the surface of the model and triangulating the nodes so that well-shaped tetrahedral elements are obtained.

- **Recursive spatial decomposition algorithm**
  This algorithm divides a solid model into octants and further subdivides each octant according to the mesh size.

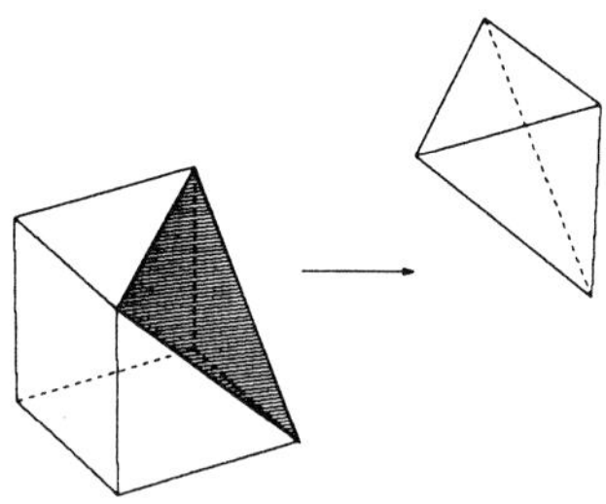

Fig. 14.1. Tetrahedron extraction from a model by introducing a planar cut

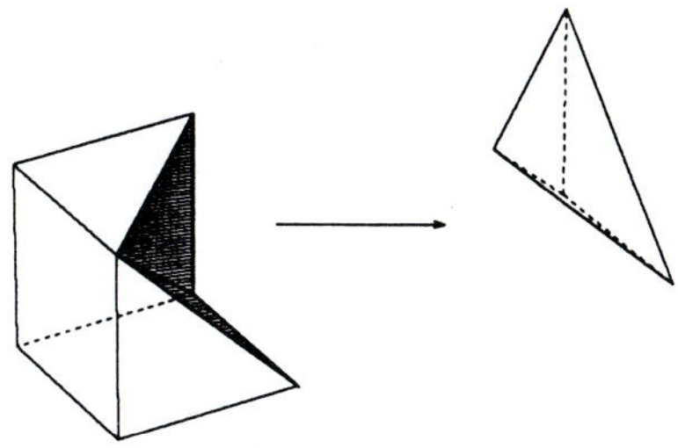

Fig. 14.2. Tetrahedron extraction from a model by introducing two planar cuts

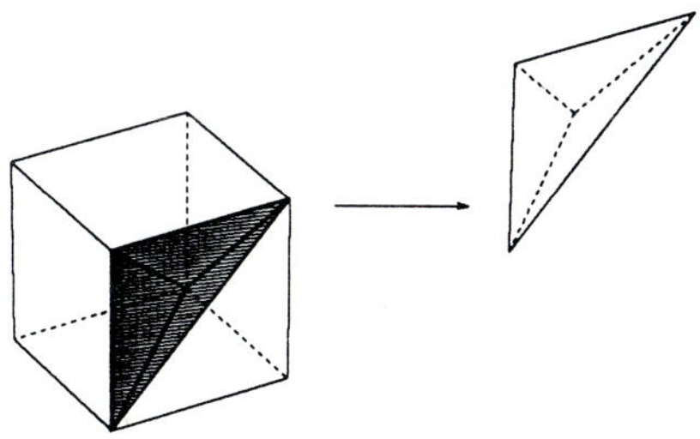

Fig. 14.3. Tetrahedron extraction from a model by introducing three planar cuts

These operations are automatic, but are not necessarily used for all kinds of models[123].

Presently the most popular method for obtaining proper meshes is known as **mapped meshing**. This method divides a model into simple-shaped hexahedral domains called **blocks** and then decomposes the model into element meshes at equal intervals. The method requires user interaction. At present, a study is being made on for automatic block decomposition by the recognition of features such as holes or protrusions[112].

The **adaptive method**, which optimizes meshes, is at present put into practical use. This method assumes the errors in the result of an analysis and optimizes the meshes, in one of the following three ways:

r-method: Optimizes meshes by displacing nodes distributed within or on the surface of a solid model. The number of divisions remains unchanged, so the amount of calculation does not increase. However optimization is limited.

h-method: Optimizes meshes by subdividing elements that contain large errors. Optimization is not limited but the amount of calculation increases.

p-method: Optimizes meshes by elevating the degree of approximating polynomial elements which contain large errors. There is no limitation to optimization, but a solver that can deal with high-degree elements is needed.

If a designer can handle these adaptive methods, then the first mesh generation need not be done by an experienced engineer.

Some solvers that perform the p-method incorporate a function to optimize shapes according the given conditions, in addition to a function to analyze the shapes[170].

It is difficult to know whether a reliable simulation can be performed. Actually, simulation does not always give the result equivalent to the one obtained in the actual experiment. If an analytic model is slightly modified, the simulation result may be largely changed. To avoid such an unstable situation, an examination is taken, in using the AI technology, to enable anyone to use the know-how of experienced engineers in mesh generation or analysis[89].

14.5 KSWAD — Integrated CAE System

This section introduces KSWAD (Kubota Super Workstation Analysis and Design) of an integrated CAE system as an example of the application of a solid modeler to a preprocessor.

14.5.1 How to apply data of solid modeler

KSWAD incorporates KSWAD/PREPOST (PREPOST from henceforth) as a pre/postprocessor for the finite element method and KSWAD/SOLID (SOLID from henceforth) as a solid modeler. DESIGNBASE is applied to SOLID, (where the basic functions follow that of DESIGNBASE) and there are extra functions for interfacing with PREPOST. PREPOST can handle the four kinds of shape data: — point, curve, surface, and solid. The data on curves and surfaces in DESIGNBASE is represented in PREPOST as shown in Table 14.1.

Shape data is constructed in DESIGNBASE, converted to data for PREPOST, and then written in an intermediate file. PREPOST receives line data for edges, face data for surfaces, and solid data for solids. Each edge, surface, and solid has the following data:

edge : start and end point IDs and geometric data of the edge
surface : boundary edge ID and geometric data of the surface
solid : boundary surface ID

As shown above, the solid data in PREPOST does not include any topological data other than the boundary surface ID.

Table 14.1. Data conversion from DESIGNBASE to PREPOST

	DESIGNBASE	KSWAD/PREPOST
curve	straight line	straight line
	arc	rational Bézier curve
	Bézier curve	rational Bézier curve
	rational Bézier curve	rational Bézier curve
surface	plane	plane
	quadric surface	RBG patch
	Bézier surface	RBG patch
	rational Bézier surface	RBG patch
	Gregory patch	RBG patch
	RBG patch	RBG patch

Note: RBG patch = rational boundary Gregory patch

SOLID incorporates the functions of DESIGNBASE that adapt data in PREPOST. It also incorporates an extra function to merge boundary surfaces. Take the example of a solid that is divided into two independent solids by a plane. The extra function regards the two sectional surfaces obtained by the division as one surface and sends the data to PREPOST. With this operation, the two independent solids are recognized as being connected and sharing a common boundary surface. This function is helpful when creating meshes from a solid of complicated shape.

PREPOST can create meshes from the shape data in SOLID, by using the algorithm of either **genuine automatic meshing** or **mapped meshing**. The characteristics of these are shown in Table 14.2.

Table 14.2. Characteristics of meshing algorithms

	genuine automatic meshing	mapped meshing
shape	any shape	hexahedron (parallelepiped) or pentahedron (wedge)
element	tetrahedron only	hexahedron, pentahedron

When using the mapped meshing algorithm, a solid should be hexahedral or pentahedral in shape; thus it should be divided beforehand by SOLID or PREPOST.

14.5.2 Genuine automatic meshing

Automatic meshing is an algorithm to automatically decompose a solid into tetrahedral elements by specifying the size of the meshes. Usually, the mesh size specified applies to the whole solid. When a local change is to be made, the size should be specified at the end point of an edge of the mesh. Figure 14.4 shows the procedure involved in genuine automatic meshing in PREPOST.

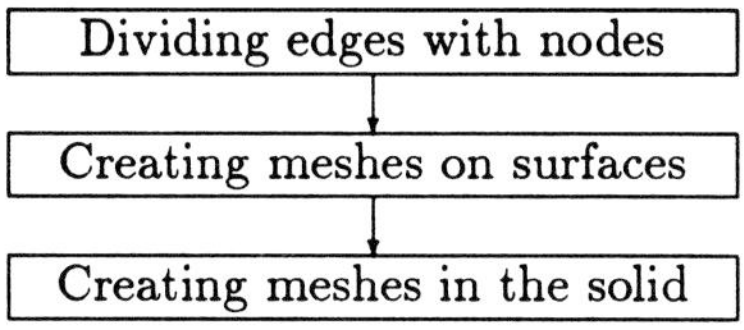

Fig. 14.4. Procedure of genuine automatic meshing

Let us consider this procedure step by step.

- Dividing edges with nodes: Distribute nodes to edges of a solid with a given mesh size. Consider local mesh sizes if specified, then divide nodes using geometric progression.

- Creating meshes on surfaces: Create meshes on the solid surfaces. The surfaces should be planar in order to create meshes. When solid data is sent to PREPOST, surfaces that are not planar are converted into rational boundary Gregory patches, then the rational boundary Gregory patches are projected onto planes. The normal vectors of the planes onto which the rational boundary Gregory patches are projected should be obtained by averaging 25 normal vectors, that is, 5 vectors for each surface with parameters u and v.

  A mesh on a plane is a polygon that consists of edges to which nodes have been distributed. Figure 14.5 provides an illustration.

  Let us visualize this step using the above figure. At first, define an interior point P_k, for the ith edge of this polygonal plane. With this point a triangle (P_{i-1}, P_i, P_k) inside the plane is defined, which is formed to be a regular triangle as far as possible. The remainder of the original plane now undergoes meshing. If the angle at P_i is very small. other triangle is created by connecting P_{i-1} and P_{i+1}, and if the angle at P_{i-1} is very small, other triangle is also created by connecting P_i and P_{i-2}. If the interior point P_k lies close to a boundary point, connect the boundary point to P_i and P_{i-1} so as to create another triangle. The choice

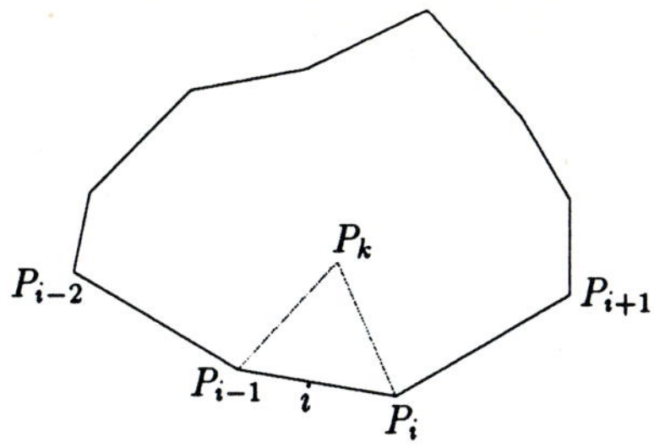

Fig. 14.5. Mesh creation on a surface

of method for creating the triangle depends on the aspect ratio of the triangle. Repeat these operations until the whole plane is divided into triangles.

Next, perform the **smoothing** for the meshes created. This operation improves the aspect ratio of the triangles created by displacing the nodes. The coordinates of an interior point P_i to which a node is to be displaced can be obtained with the following equation:

$$P_i = \frac{P_1 C_1 + \cdots + P_n C_n}{C_1 + \cdots + C_n}$$

where $P_j(j = 1 \cdots n)$ indicates the coordinates of a point connected to the interior point P_i, and $C_j(j = 1 \cdots n)$ indicates the weight of P_j. These values depend on the mesh size.

The mesh size l_j with the interior point P_j is obtained as follows (Figure 14.6):

$$l_j = \frac{\sum_{k=1}^{m} \frac{l_k^2}{r_k}}{\sum_{k=1}^{m} \frac{l_k}{r_k}}$$

where m indicates the division number for the boundary edges, r_k indicates the distance between the kth boundary edge and P_j, and l_k indicates the mesh size of the kth boundary edge. The weight C_j at P_j is obtained from the mesh size l_j as

$$C_j = \frac{1}{l_j}$$

where $C_j = 1$ if the sizes of the adjacent meshes are the same.

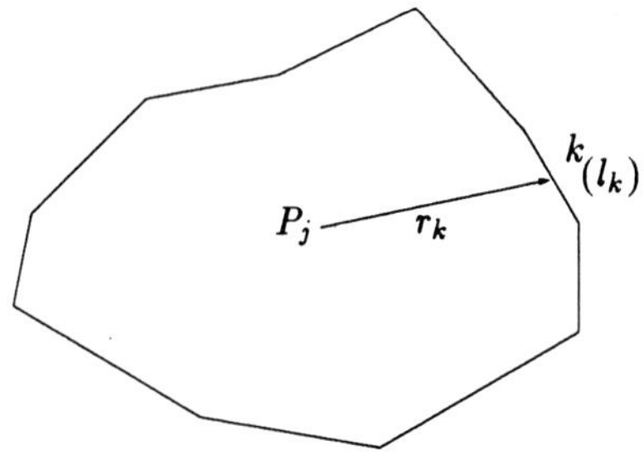

Fig. 14.6. Mesh size of an interior point

- Creating meshes in the solid: When all the boundary surfaces of a solid are decomposed into meshes, the solid is a closed interval area covered with triangular meshes. In the same way as for mesh creation on surfaces, define an interior point P_k about the ith triangular mesh on a boundary surface of a solid (see Figure 14.7), so that a tetrahedron $(P_{i1}, P_{i2}, P_{i3}, P_k)$ is created.

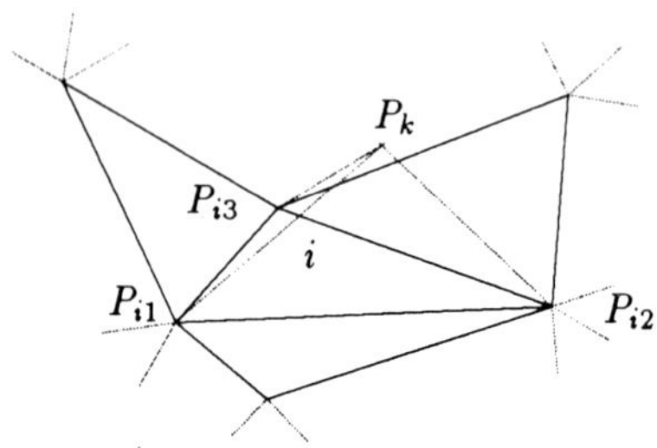

Fig. 14.7. Mesh creation in a solid

The interior point P_k should be chosen so that the tetrahedron resembles a regular tetrahedron as far as possible. Then, new boundary surfaces which include the newly-created triangular meshes (P_{i1}, P_{i2}, P_k), (P_{i2}, P_{i3}, P_k), and (P_{i3}, P_{i1}, P_k) are created. If the interior angles between the ith mesh and its adjacent meshes are considerably small, then other tetrahedrons are defined by these meshes. If the interior point P_k lies near to a boundary point, then connect the points so as to create another tetrahedron. Repeat these operations until the closed interval area is filled with tetrahedral meshes. Finally perform the smoothing of the meshes created, as is done for surface meshes.

14.5.3 Mapped meshing

Mapped meshing is a method for creating nodes and meshes on a surface or in a solid by the interpolation of edges and surfaces that construct the solid. In general, when mapped meshing is performed for a surface, the actual surface data is first converted to fit to the normalized 2-dimensional coordinate system, and meshes are created on the converted surface. Then, the data is reverted to the original surface. Nodes and meshes of a solid can also be created through a normalized 3-dimensional coordinate system, analogous to the operation for a surface.

With PREPOST, meshes can be created for a surface with three or four sides and for a solid with five or six faces (prism and parallelepiped, respectively).

- Algorithm

  The algorithm for mapped meshing consists of the following three major operations:

  1. Create a function that defines the relationship between the actual geometric data and the normalized coordinate system.
  2. Create nodes and meshes on a normalized surface or in a normalized solid.
  3. Invert the function created in the first step and return the nodes to the original surface or solid.

  Let us now consider these operations in detail.

  At first, define a ξ-η-γ coordinate system, the origin of which is one of the vertices of a parallelepiped and whose axes coincide with the edges of the parallelepiped (see Figure 14.8). Number each side of the parallelepiped, using the following procedure:

1. $(\eta = 0, \gamma = 0) \quad x = f_1(\xi), y = g_1(\xi), z = h_1(\xi)$
2. $(\eta = 0, \gamma = 1) \quad x = f_2(\xi), y = g_2(\xi), z = h_2(\xi)$
3. $(\eta = 1, \gamma = 1) \quad x = f_3(\xi), y = g_3(\xi), z = h_3(\xi)$
4. $(\eta = 1, \gamma = 0) \quad x = f_4(\xi), y = g_4(\xi), z = h_4(\xi)$
5. $(\xi = 0, \gamma = 0) \quad x = f_5(\eta), y = g_5(\eta), z = h_5(\eta)$
6. $(\xi = 0, \gamma = 1) \quad x = f_6(\eta), y = g_6(\eta), z = h_6(\eta)$
7. $(\xi = 1, \gamma = 1) \quad x = f_7(\eta), y = g_7(\eta), z = h_7(\eta)$
8. $(\xi = 1, \gamma = 0) \quad x = f_8(\eta), y = g_8(\eta), z = h_8(\eta)$
9. $(\xi = 0, \eta = 0) \quad x = f_9(\gamma), y = g_9(\gamma), z = h_9(\gamma)$
10. $(\xi = 0, \eta = 1) \quad x = f_{10}(\gamma), y = g_{10}(\gamma), z = h_{10}(\gamma)$
11. $(\xi = 1, \eta = 1) \quad x = f_{11}(\gamma), y = g_{11}(\gamma), z = h_{11}(\gamma)$
12. $(\xi = 1, \eta = 0) \quad x = f_{12}(\gamma), y = g_{12}(\gamma), z = h_{12}(\gamma)$

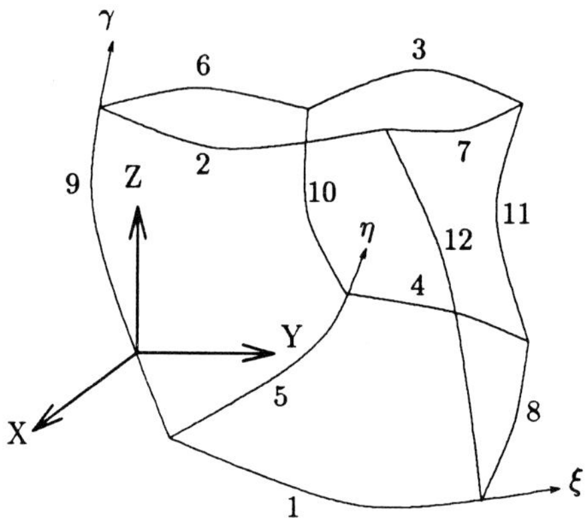

Fig. 14.8. ξ-η-γ coordinate system

This points of Edge 1 are defined by $\eta = 0, \gamma = 0$ and the coordinates x, y, and z can be defined by the functions f_1, g_1, and h_1 of ξ. ξ, η, and γ of the start and end points of Side 1 can also be expressed by (0,0,0) and (1,0,0). In PREPOST, f_1, g_1, and h_1 are approximated with polylines depending on the original shape which is adjusted to the distance from the starting point. Obtain the functions f_i, g_i, and h_i for all edges from 1 to 12 as well.

Next, according to the input data on the division number, division ratio, and mesh type of each side, create nodes and meshes in the ξ-η-γ coordinate system. At this stage, the node coordinates are defined by (ξ, η, γ). By using the previously defined expressions of f_i, g_i, and $h_i (1 \leq i \leq 12)$ together with the following transformation equation, convert the ξ, η, γ coordinates of the nodes created into x, y, z coordinates:

$$\begin{aligned} x(\xi,\eta,\gamma) &= (1-\eta)(1-\gamma)f_1(\xi) + (1-\eta)\gamma f_2(\xi) \\ &\quad +\eta\gamma f_3(\xi) + \eta(1-\gamma)f_4(\xi) \\ &\quad +(1-\xi)(1-\gamma)f_5(\eta) + (1-\xi)\gamma f_6(\eta) \\ &\quad +\xi\gamma f_7(\eta) + \xi(1-\gamma)f_8(\eta) \\ &\quad +(1-\xi)(1-\eta)f_9(\gamma) + (1-\xi)\eta f_{10}(\gamma) \\ &\quad +\xi\eta f_{11}(\gamma) + \xi(1-\eta)f_{12}(\gamma) \\ &\quad +c(\xi,\eta,\gamma) \end{aligned}$$

where

$$\begin{aligned} c(\xi,\eta,\gamma) &= -2\{(1-\xi)(1-\eta)(1-\gamma)x(0,0,0) \\ &\quad +(1-\xi)(1-\eta)\gamma x(0,0,1) \\ &\quad +(1-\xi)\eta(1-\gamma)x(0,1,0) \end{aligned}$$

$$
\begin{aligned}
&+(1-\xi)\eta\gamma x(0,1,1)\\
&+\xi(1-\eta)(1-\gamma)x(1,0,0)\\
&+\xi(1-\eta)\gamma x(1,0,1)\\
&+\xi\eta(1-\gamma)x(1,1,0)\\
&+\xi\eta\gamma x(1,1,1)\}
\end{aligned}
$$

In the same way as above, the coordinates y and z can be obtained by using the functions g_i and h_i. If the geometric shape of a surface cannot be represented correctly by interpolating with the defining edges, then the expression for transforming the shape according to the geometric data on the surface should be defined when the coordinate systems are converted.

- **Procedure for mapped meshing**

1. Editing a solid: To perform mapped meshing, a solid should be a pentahedron or hexahedron. Therefore, other types of solids should first be divided by using SOLID or PREPOST. To maintain the continuity of meshes, the boundaries of such solids are required to be defined by a common surface.
2. Specifying the division number and division ratio of each side: Specify the division number and division ratio of the boundary sides in order to create points inside the boundaries. Geometric progression, arithmetic progression, or random input can be used. If two solids share their boundaries and nodes are already created, then there is no need to specify the division number.
3. Executing the mapped meshing of a solid: Input the mesh type. Meshes are then created inside of the solid.

14.5.4 Examples of analyses by KSWAD

Plate 7 in the frontispiece shows the distribution of principal stress in one quarter of a connecting rod. Automatic meshing is used to create these meshes. Plate 8 shows the results of the analysis on frame material. The model on the right side shows its deformation and the one on the left side shows the stress. Mapped meshing is used after dividing the solid into ten parallelepipeds using the cutting function.

References

[1]J.H. Argyris and S. Kelsey. *Energy Theorems and Structural Analysis.* Butterworth and Company, 1960.

[2]D.B. Arnold and D.A. Duce. *ISO Standards for Computer Graphics.* Butterworths, 1990.

[3]R.E. Barnhill, J.H. Brown, and I.M. Klucewicz. A new twist in computer aided geometric design. *Computer Graphics and Image Processing*, 8:78–91, 1978.

[4]R.E. Barnhill, G. Farin, M. Jordan, and B.R. Piper. Surface/surface intersection. *Computer Aided Geometric Design*, 4(1–3):3–16, 1987.

[5]R.E. Barnhill and S.N. Kersey. A marching method for parametric surface/surface intersection. *Computer Aided Geometric Design*, 7(1):257–280, 1990.

[6]B.A. Barsky. *The Beta-Spline: A Local Representation Based on Shape Parameters and Fundamental Geometric Measures.* PhD thesis, University of Utah, Salt Lake City, Utah, 1981.

[7]B.G. Baumgart. Geometric modelling for computer vision. Technical report, Report STAN-CS-74-463, Stanford University: Stanford Artificial Intelligence Laboratory, 1974.

[8]B.G. Baumgart. A Polyhedron Representation for Computer Vision. In *AFIPS Proceedings*, volume 44, pages 589–596, 1975.

[9]P.J. Besl and R.C. Jain. Three-dimensional object recognition. *ACM Computing Surveys*, 17(1):75–145, 1985.

[10]Bézier,P.o. *Numerical Control: Mathematics and Applications.* John Wiley & Sons, London, 1972.

[11]Bézier,P.o. Définition numérique des courbes et surfaces. *Automatisme*, 11:625–632, 1966.

[12]Bézier,P.n. *The Mathematical Basis of the UNISURF CAD System.* Butterworths, London, 1986.

[13]Bézier,P.n. Définition numérique des courbes et surfaces(ii). *Automatisme*, 12:17–21, 1967.

[14]J.F. Blinn. Models of light reflection for computer synthesized pictures. *Computer Graphics*, 11(2):192–198, 1977.

[15]W. Boehm. Curvature continuous curves and surfaces. *Computer Aided Geometric Design*, 2(4):313–323, 1985.

[16]J.W. Boyse. Interference detection among solids and surfaces. *Communications of the ACM*, 22(1):3–9, 1979.

[17]J.W. Boyse and J.E. Gilchrist. GMSolid: Interactive modeling for design and analysis of solids. *IEEE Computer Graphics and Applications*, 2(3):27–39, 1982.

[18]I.C. Braid, R.C. Hillyard, and I.A. Stroud. Stepwise construction of polyhedra in geometric modelling. In K.W. Brodlie, editor, *Mathematical Methods in Computer Graphics and Design*, pages 123–141. Academic Press, 1980.

[19]I.C. Braid and C.A. Lang. Computer-aided design of mechanical components with volume building bricks. In *Proceedings of PROLAMAT '73*, 1973.

[20]C.M. Brown. PADL-2: A technical summary. *IEEE Computer Graphics and Applications*, 2(2):69–84, 1982.

[21]S.A. Brown, C.E. Drayton, and B. Mittman. A description of the APT language. *Communications of the ACM*, 6(11), 1963.

[22]W.S Burnside and A.W. Panton. *The Theory of Equation.* Longmans, Green and Co., London, 3rd edition, 1892.

[23]I. Carlbom, I. Chakravarty, and D. Vanderschel. A hierarchical data structure for representing the spatial decomposition of 3-D objects. *IEEE Computer Graphics and Applications*, 5(4):24–31, 1985.

[24]L. Carpenter. The A-buffer, an antialiased hidden surface method. *Computer Graphics*, 18(3):103–108, 1984.

[25]M.S. Casale. Free-form solid modeling with trimmed surface patches. *IEEE Computer Graphics and Applications*, 7(1):33–43, 1987.

[26]M.S. Casale and J.E. Bobrow. A set operation algorithm for sculptured solids modeled with trimmed patches. *Computer Aided Geometric Design*, 6(3):235–247, 1989.

[27]E. Catmull. *A Subdivision Algorithm for Computer Display of Curved Surfaces.* PhD thesis, University of Utah, 1974.

[28]J.J. Chen and T.M. Ozsoy. Predictor-corrector type of intersection algorithm for C^2 parametric surfaces. *Computer Aided Design*, 20(6):347–352, 1988.

[29]K.-P. Cheng. Using plane vector fields to obtain all the intersection curves of two general surfaces. In W. Strasser and H.P. Seidel, editors, *Theory and Practice of Geometric Modeling*, pages 187–204. Springer-Verlag, Berlin, 1989.

[30]H. Chiyokura. *A study of modelling solids with free-form surfaces.* PhD thesis, University of Tokyo, 1983. (in Japanese).

[31]H Chiyokura. *Solid Modeling.* Kogyo Chousakai, 1985. (in Japanese).

[32]H. Chiyokura. Localized surface interpolation for irregular meshes. In T.L. Kunii, editor, *Advanced Computer Graphics - Proceedings of Computer Graphics Tokyo 86*, pages 289–298, Tokyo, 1986. Springer-Verlag.

[33]H. Chiyokura. An extended rounding operation for modeling solids with free-form surfaces. *IEEE Computer Graphics and Applications*, 7(12):27–36, 1987.

[34]H. Chiyokura. *Solid Modelling with DESIGNBASE: Theory and Implementation.* Addison-Wesley, Reading, Massachusetts, 1988.

[35]H. Chiyokura and F. Kimura. Design of solids with free-form surfaces. *Computer Graphics*, 17(3):289–298, 1983.

[36]H. Chiyokura, T. Takamura, K. Konno, and T. Harada. G^1 surface interpolation over irregular meshes with rational curves. In G. Farin, editor, *Frontiers in Geometric Modeling.* SIAM, Philadelphia, 1990.

[37]B.K. Choi and S.Y. Ju. Constant-radius blending in surface modelling. *Computer Aided Design*, 21(4):213–220, 1989.

[38]J. Chung and M.D. Schuessel. Comparison of variational and parametric design. In *AUTOFACT '89 Conference Proceedings*, pages 5_27–5_44, Dearborn, Michigan, 1989. Society of Manufacturing Engineers.

[39]R.W. Clough. The finite element method in plane stress analysis. In *Proceedings of 2nd ASCE Conference on Electronic Computation*, 1960.

[40]E. Cohen, T. Lyche, and R.F. Riesenfeld. Discrete B-splines and subdivision techniques in computer aided geometric design and computer graphics. *Computer Graphics and Image Processing*, 14(2):87–111, 1980.

[41]M.F. Cohen, S.E. Chen, J.R. Wallace, and D.P. Greenberg. A progressive refinement approach to fast radiosity image generation. *Computer Graphics*, 22(4):75–84, 1988.

[42]PHIGS+ Committee. PHIGS+ functional description revision 3.0. *Computer Graphics*, 22(3):125–218, 1988.

[43]R.L. Cook and K.E. Torrance. A reflectance model for computer graphics. *Computer Graphics*, 15(3):307–316, 1981.

[44]S.A. Coons. *Surfaces for Computer-Aided Design of Space Figures*. MIT, 1964.

[45]S.A. Coons. *Surfaces for Computer-Aided Design of Space Forms*. MIT, 1967.

[46]M.G. Cox. The numerical evaluation of B-splines. DNAC 4, National Physical Laboratory, 1971.

[47]C. de Boor. On calculating with B-splines. *Journal of Approximation Theory*, 6:50–62, 1972.

[48]P. de Casteljau. Outillages méthodes calcul. Technical report, A. Citroen, Paris, 1959.

[49]P. de Casteljau. Courbes et surface à poles. Technical report, A. Citroen, Paris, 1963.

[50]T. Dokken. Finding intersection of B-spline represented geometries using recursive subdivision techniques. *Computer Aided Geometric Design*, 2(1–3):189–195, 1985.

[51]T. Dokken and Dæhlen,M. and Lyche,T. and Mørken,T. Good approximation of circles by curvature-continuous Bézier curves. *Computer Aided Geometric Design*, 7(1–4):33–41, 1990.

[52]G. Enderle, K. Kansy, and G. Pfaff. *Computer Graphics Programming. GKS-The Graphics Standard*. Springer-Verlag, 1984.

[53]G. Farin. *Curves and Surfaces for Computer Aided Geometric Design: A Practical Guide*. Academic Press, San Diego, CA, 1988.

[54]R.T. Farouki. Direct surface section evaluation. In G. Farin, editor, *Geometric Modeling: Algorithm and New Trends*, pages 319–334. SIAM, Philadelphia, PA, 1987.

[55]J.D. Foley, A. van Dam, S.K. Feiner, and J.H. Hughes. *Computer Graphics Principles and Practice*. Addison-Wesley, second edition, 1990.

[56]K. Fujimura, H. Toriya, K. Yamaguchi, and T.L. Kunii. Octree algorithms for solid modeling. In T.L. Kunii, editor, *Computer Graphics - Theory and Application*, pages 96–110. Springer-Verlag, Tokyo, 1983.

[57]R.N. Goldman, T.W. Sederberg, and D.C. Anderson. Vector elimination. *Computer Aided Geometric Design*, 1:327–356, 1984.

[58]C.M. Goral, K.E. Torrance, D.P. Greenberg, and B. Battaile. Modeling the interaction of light between diffuse surfaces. *Computer Graphics*, 18(3):213–222, 1984.

[59]W.J. Gordon and R.F. Riesenfeld. Bernstein-Bézier methods for the computer aided design of free-form curves and surfaces. *Journal of the ACM*, 21:293–310, 1974.

[60]H. Gouraud. Continuous shading of curved surfaces. *IEEE Transactions on Computers*, 20(6):623–628, 1971.

[61]A.R. Grayer. The automatic production of machined components stating from a stored geometric description. In D. McPherson, editor, *Advanced Computer-Aided Manufacturing*, pages 137–151. North-Holland, Amsterdam, 1977.

[62]J.A. Gregory. Smooth interpolation without twist constraints. In R.E Barnhill and R.F Riesenfeld, editors, *Computer Aided Geometric Design*, pages 71–87. Academic Press, 1974.

[63]T. Harada, H. Toriya, and H. Chiyokura. An enhanced rounding operation between curved surfaces in solid modeling. In T.S. Chua and T.L. Kunii, editors, *CG International '90*, pages 563–588. Springer-Verlag, Tokyo, 1990.

[64]K. Ho-Le. Finite element mesh generation methods: A review and classification. *Computer Aided Design*, 20(1):27–38, 1988.

[65]C.M. Hoffmann. *Geometric & Solid Modeling, An Introduction.* Morgan Kaufmann, San Meteo, CA, 1989.

[66]Holmström,L. and Laakko,T. Rounding facility for solid modelling of mechanical parts. *Computer Aided Design*, 20(10):605–614, 1988.

[67]F.R.A. Hopgood, D.A. Duce, J.R. Gallop, and D.C. Sutcliffe. *Introduction to the Graphical Kernel System(GKS).* Academic Press, London, 1983.

[68]J. Horgan, R. Schmidtberg, and M. Yerry. The marriage between solid modeling and finite element technologies. In *Fourth International Solid Modeling Conference*, 1987.

[69]E.G. Houghton, R.F. Emnett, J.D. Factor, and C.L. Sabharwal. Implementation of a divide-and-conquer method for intersection of parametric surfaces. *Computer Aided Geometric Design*, 2(1–3):173–183, 1985.

[70]A Hrenikoff. Solution of problems in elasticity by the framework method. *J. Appl. Mech.*, 8, 1941.

[71]M Imai and Sugiyama N. *VLSI design silicon compilation.* Science Forum, Tokyo, 1988.

[72]D.S. Immel, M.F. Chen, and D.P. Greenberg. A radiosity method for non-diffuse environments. *Computer Graphics*, 20(4):133–142, 1986.

[73]International Standards Organization. *Graphical Kernel System, ISO 7492*, 1985.

[74]International Standards Organization. *Graphical Kernel System for Three Dimensions(GKS-3D), ISO/DIS 8805*, 1987.

[75]International Standards Organization. *External Representation of Product Definition Data, ISO Draft Proposal 10303, ISO TC184/SC4 N38*, 1989.

[76]C.L. Jackins and S.L. Tanimoto. Oct-trees and their use in representing three-dimensional objects. *Computer Graphics and Image Processing*, 4:249–270, 1980.

[77]R. Johnson. Reaching new heights with solid modeling. *Computer Graphics World*, 12(12):65–72, 1989.

[78]J.T. Kajiya. The rendering equation. *Computer Graphics*, 20(4):143–150, 1986.

[79]S. Kawabe. *Next generation CAD/CAM(in Japanese).* Kogyo Chousakai, 1985.

[80]P.A. Koparkar and S.P. Mudur. Generation of continuous smooth curves resulting from operations on parametric surface patches. *Computer Aided Design*, 18(4):193–206, 1986.

[81]T.L. Kunii, I. Fujishiro, and X. Mao. G-quadtree: A hierarchical representation of gray-scale digital images. *The Visual Computer*, 2(4):219–226, 1986.

[82]J.M. Lane and R.F. Riesenfeld. A theoretical development for the computer generation and display of piecewise polynomial surfaces. *IEEE Transactions on Pattern Analysis and Machine Intelligence*, 2(1):35–46, 1980.

[83]E.T.Y. Lee. The rational Bézier representation for conics. In G. Farin, editor, *Geometric Modeling: Algorithm and New Trends*, pages 3–19. SIAM, Philadelphia, PA, 1987.

[84]Y.T. Lee and A.A.G. Requicha. Algorithms for computing the volume and other integral properties of solids: I. - Known methods and open issues. *Communications of the ACM*, 25(9):635–641, 1982.

[85]Y.T. Lee and A.A.G. Requicha. Algorithms for computing the volume and other integral properties of solids: II. - A family of algorithms based on representation conversion and cellular approximation. *Communications of the ACM*, 25(9):642–650, 1982.

[86]J. Levin. A parametric algorithm for drawing pictures of solid objects composed of quadric surfaces. *Communications of the ACM*, 19(10):555–563, 1976.

[87]J. Levin. Mathematical models for determining the intersections of quadric surfaces. *Computer Graphics and Image Processing*, 11:73–87, 1979.

[88]S. Lien and J.T. Kajiya. A symbolic method for calculating the integral properties of arbitrary nonconvex polyhedra. *IEEE Computer Graphics and Applications*, 4(10):35–41, 1984.

[89]Y Maekawa. Possibilities in mechanical cae. *CAD & CIM*, 4(2):48–52, 1990. (in Japanese).

[90]M. Mäntylä. *An Introduction to Solid Modeling.* Computer Science Press, Maryland, 1988.

[91]Mäntylä,M. and Sulonen,R. GWB: A solid modeler with Euler operators. *IEEE Computer Graphics and Applications*, 2(7):17–31, 1982.

[92]Mäntylä,M. and Takala,T. The geometric workbench(GWB) - an experiment geometric modeling system. In *EUROGRAPHICS 81*, pages 205–215, North-Holland, Amsterdam, 1981.

[93]R.P. Markot and R.L. Magedson. Solutions of tangential surface and curve intersections. *Computer Aided Design*, 21(7):421–427, 1989.

[94]R.R. Martin. *Principal Patches for Computational Geometry.* PhD thesis, Cambridge University Engineering Department, 1982.

[95]K. Matsushima. *yasashii CAD/CAM.* Kougyou Chousakai, 1983. (in Japanese).

[96]D. Meagher. Geometric modeling using octree encoding. *Computer Graphics and Image Processing*, 19:129–147, 1982.

[97]J.R. Miller. Geometric approaches to nonplanar quadric surface intersection curves. *ACM Transactions on Graphics*, 6(4):274–307, 1987.

[98]I. Misumi, I. Nomura, H. Nagai, and H. Yokoi. Mass property calculation in solid modeling using green's formula. *The Spring Conference Proceedings of the Japan Society of Precision Engineering*, pages 655–656, 1987. (in Japanese).

[99]T. Nishita, K. Kaneda, and E. Nakamae. High-quality rendering of parametric surfaces by using robust scanline algorithm. In T.S. Chua and T.L. Kunii, editors, *CG International '90*, pages 493–506, Tokyo, 1990. Springer-Verlag.

[100]T. Nishita and E. Nakamae. Scan conversion of regions bounded by bézier curves - application to outline fonts. *Reports on Graphics and CAD of Information Processing Society of Japan*, 45(2):1–8, 1990. (in Japanese).

[101]T. Nishita, T. Sederberg, and M. Kakimoto. Ray tracing trimmed rational surface patches. *Computer Graphics*, 24(4):337–345, 1990.

[102]K. Noborio, S. Fukuda, and T. Arimoto. Transformation algorithm from b-reps to octree data and its evaluation. *Transactions of Information Processing Society of Japan*, 28(10):1003–1012, 1987. (in Japanese).

[103]N. Okino, Y. Kakazu, and H. Kubo. TIPS-1: Technical information processing system for computer-aided design, drawing and manufacturing. In J. Hatvany, editor, *Computer Languages for Numerical Control*, pages 141–150. North-Holland, Amsterdam, 1973.

[104]D.R. Peachy. Solid texturing of complex surfaces. *Computer Graphics*, 19(3):279–286, 1985.

[105]Q.S. Peng. An algorithm for finding the intersection line between two B-spline surfaces. *Computer Aided Design*, 16(4):191–196, 1984.

[106]K. Perlin. An image synthesizer. *Computer Graphics*, 19(3):287–296, 1985.

[107]D.P. Perterson. Boundary to constructive solid geometry mapping: A focus on 2D issues. *Computer Aided Design*, 18(1):3–14, 1986.

[108]B.T. Phong. Illumination for computer generated images. *Communications of the ACM*, 13(6):311–317, 1975.

[109]M.S. Pickett, R.B. Tilove, and V. Shapiro. Roboteach: An off-line robot programming system. In M.S. Pickett and J.W. Boyse, editors, *Solid Modeling by Computers*, pages 159–184. Plenum Press, New York, 1984.

[110]L. Piegl and W. Tiller. Curve and surface constructions using rational B-splines. *Computer Aided Design*, 19(9):485–498, 1987.

[111]M.J. Pratt and A.D. Geisow. Surface/surface intersection problems. In J.A. Gregory, editor, *The Mathematics of Surfaces*, pages 117–142. Oxford University Press, 1986.

[112]A. Razdan, M.R. Henderson, P.F. Chavez, and P.A. Erickson. Feature based object decomposition for finite element meshing. *Visual Computer*, 5(5):291–303, 1989.

[113]A.A.G. Requicha. Representation for rigid solids: Theory, method and systems. *ACM Computing Surveys*, 12(4):437–464, 1980.

[114]A.A.G. Requicha and H.B. Voelcker. Solid modeling: A historical summary and contemporary assessment. *IEEE Computer Graphics and Applications*, 2(2):9–24, 1982.

[115]A.A.G. Requicha and H.B. Voelcker. Boolean operations in solid modeling: Boundary evaluation and merging algorithms. *Proceedings of the IEEE*, 73(1):30–44, 1985.

[116]R.F. Riesenfeld. *Application of B-spline Approximation to Geometric Problems of Computer Aided Design*. PhD thesis, Syracuse University, Syracuse, N.Y., 1972.

[117]A.P. Rockwood and J.C. Owen. Blending surfaces in solid modeling. In G.E. Farin, editor, *Geometric Modeling: Algorithms and New Trends*, pages 367–383. SIAM, Philadelphia, PA, 1987.

[118]D.F. Rogers and J.A. Adams. *Mathematical Elements for Computer Graphics*. McGraw-Hill, second edition, 1990.

[119]J.R. Rossignac and A.A.G. Requicha. Constant-radius blending in solid modelling. *Computers in Mechanical Engineering*, 3:65–73, 1984.

[120]J.R. Rossignac and A.A.G. Requicha. Offsetting operations in solid modeling. *Computer Aided Geometric Design*, 3(2):129–148, 1986.

[121]J.R. Rossignac and H.B. Voelcker. Active zones in CSG for accelerating boundary evaluation, redundancy elimination, interference detection, and shading algorithms. *ACM Transactions on Graphics*, 8(1):51–87, 1989.

[122]R.J. Rost. *PEX Introduction and Overview, PEX Version 3.20.* MIT X Consortium, 1988.

[123]N. Sapidis and R. Perucchio. Advanced techniques for automatic finite element meshing from solid models. *Computer Aided Design*, 21(4):248–253, 1989.

[124]R.F. Sarraga. Algebraic methods for intersections of quadric surfaces in GM-Solid. *Computer Vision, Graphics, and Image Processing*, 22:222–238, 1983.

[125]R.F. Sarraga and W.C. Waters. Free-form surfaces in GMSolid: Goals and issues. In S. Pickett, Mary and W. Boyse, John, editors, *Solid Modeling by Computers from Theory to Applications*, pages 187–209. Plenum Press, New York, 1984.

[126]T.W. Sederberg. *Implicit and Parametric Curves and Surfaces for Computer Aided Geometric Design.* PhD thesis, Purdue University, 1983.

[127]T.W. Sederberg and D.C. Anderson. Ray tracing of Steiner patches. *Computer Graphics*, 18(3):159–164, 1984.

[128]T.W. Sederberg, D.C. Anderson, and R.N. Goldman. Implicit representation of parametric curves and surfaces. *Computer Vision, Graphics, and Image Processing*, 28:72–84, 1984.

[129]T.W. Sederberg, H.N. Christiansen, and S. Katz. Improved test for closed loops in surface intersections. *Computer Aided Design*, 21(8):505–508, 1989.

[130]T.W. Sederberg and S.R. Parry. Comparison of three curve intersection algorithms. *Computer Aided Design*, 18(1):58–63, 1986.

[131]T.W. Sederberg and X. Wang. Rational hodographs. *Computer Aided Geometric Design*, 4(4):333–335, 1987.

[132]M.S. Shephard, K.R. Grice, J.A. Lo, and W.J. Schroeder. Trends in automatic three-dimensional mesh generation. *Computers and Structures*, 30(1):421–429, 1988.

[133]I.J. Shoenberg. Contributions to the problem of approximation of equidistance data by analytic functions. *Quarterly of Applied Mathematics*, 4:45–99, 1946.

[134]D Shuey. PHIGS: A graphics platform for cad application development. *Computer Aided Design*, 19(8):410–417, 1987.

[135]D. Shuey, D Bailey, and T.P. Morrissey. PHIGS: A standard, dynamic, interactive graphics interface. *IEEE Computer Graphics and Applications*, 6(8):50–57, 1986.

[136]B. Smith, G.R. Rinaudot, K.A. Reed, and T. Wright. *Initial Graphics Exchange Specification(IGES) Version 4.0.* U.S. Department of Commerce, 1988.

[137]J.S. Snoeyink. Intersecting trimmed quadric surface patches: A geometric method using parametric functions. (unpublished), 1990.

[138]T.W. Stacey and A.E. Middleditch. The geometry of machining for computer-aided manufacturing. *Robotica*, 4:83–91, 1986.

[139]G. Stute. *EXAPT, Möglichkeiten und Anwendung der automatisierten Programmierung für NC Maschinen.* Carl Hanser, Munich, 1969.

[140]I.E. Sutherland. SKETCHPAD: A man-machine graphical communication system. In *Proceedings of the Spring Joint Computer Conference*, Baltimore, 1963. Spartan Books.

[141]I.E. Sutherland. Ten unsolved problems in computer graphics. *Datamation*, 12(5):22, 1966.

[142]T. Takamura, M. Ohta, H. Toriya, and H. Chiyokura. A method to convert a Gregory patch and a rational boundary Gregory patch to a rational Bézier patch and its applications. In T.S. Chua and T.L. Kunii, editors, *CG International '90*, pages 543–562, Tokyo, 1990. Springer-Verlag.

[143]K. Tei and H. Toriya. A generation method for swept solids along free-form curves. *Proceedings of JSPE Autumn Meeting in 1990*, pages 477–478, 1990. (in Japanese).

[144]K. Tei and H. Toriya. Generation of a skinning body by the sweep method. *Proceedings of the 40th Annual Convention Information Processing Society of Japan*, pages 1520–1521, 1990. (in Japanese).

[145]W. Tiller. Rational B-splines for curve and surface representation. *IEEE Computer Graphics and Applications*, 3(6), 1983.

[146]H. Togawa. *An Outline of Finite Element Methods.* baifukan, 1981. (in Japanese).

[147]H. Toriya. *A Study of an Interactive Solid Modeler Containing Free-Form Surfaces.* PhD thesis, University of Tokyo, 1989.

[148]H. Toriya, T. Satoh, K. Ueda, and H. Chiyokura. Undo and redo operations for solid modeling. *IEEE Computer Graphics and Applications*, 6(4):35–42, 1986.

[149]H. Toriya, T. Takamura, T. Satoh, and H. Chiyokura. Boolean operations of solids with free-form surfaces through polyhedral approximation. In *Proceedings of CG International '89*, pages 405–420, 1989.

[150]M.J. Turner, R.W. Clough, H.C. Martin, and L.J. Topp. Stiffness and deflection analysis of complex structures. *Journal of Aeronautical of Science*, 23(9):805–823, 1956.

[151]Várady,T. Rolling ball blends in solid modelling. In F. Kimura and Rolstadås,A., editors, *Computer Applications in Production and Engineering*, pages 295–308. North-Holland, Amsterdam, 1989.

[152]K.J. Versprille. *Computer-Aided Design Applications of the Rational B-spline Approximation Form.* PhD thesis, Syracuse University, Syracuse, N.Y., 1975.

[153]H.B. Voelcker and A.A.G. Requicha. Geometric modelling of mechanical parts and processes. *IEEE Computer*, 10(12):48–57, 1977.

[154]D.L. Vosseler. Sweep-to-CSG conversion using pattern recognition techniques. *IEEE Computer Graphics and Applications*, 5(8):61–68, 1985.

[155]G. Wang. The subdivision method for finding the intersection between two Bézier curves or surfaces. Technical report, Zhejiang University Journal, 1984.

[156]W.P. Wang and K.K. Wang. Geometric modeling for swept volume of moving solids. *IEEE Computer Graphics and Applications*, 6(12):8–17, 1986.

[157]K. Washizu, H. Miyamoto, Y. Yamada, Y. Yamamoto, and T. Kawai, editors. *Finite Element Method Handbook.* Baifukan, 1981. (in Japanese).

[158]K.J. Weiler. Edge-based data structures for solid modeling in curved-surface environments. *IEEE Computer Graphics and Applications*, 5(1):21–40, 1985.

[159]K.J. Weiler. *Topological Structures for Geometric Modeling.* PhD thesis, Rensselar Polytechnic Institute, 1986.

[160]M.A. Wesley, T. Lazano-Perez, L.I. Lieberman, M.A. Lavin, and D.D. Grossman. A geometric modelling system for automated mechanical assembly. *IBM Journal of Research and Development*, 24:64–74, 1980.

[161]T. Whitted. An improved illumination model for shaded display. *Communications of the ACM*, 23(6):343–349, 1980.

[162]P.R. Wilson. Euler formulas and geometric modeling. *IEEE Computer Graphics and Applications*, 5(8):24–36, 1985.

[163]T.C. Woo. Computer-aided recognition of volumetric design-CARVD. In D. McPherson, editor, *Advanced Computer-Aided Manufacturing*, pages 121–135. North-Holland, Amsterdam, 1977.

[164]T.C. Woo. A combinatorial analysis of boundary data structure schemata. *IEEE Computer Graphics and Applications*, 5(3):19–27, 1985.

[165]T.C. Woo and T. Thomasma. An algorithm for generating solid elements in objects with holes. *Computers and Structures*, 18(2):333–342, 1984.

[166]Wördenweber,B. Finite element mesh generation. *Computer Aided Design*, 16(5):285–291, 1984.

[167]G. Wyvill, T.L. Kunii, and Y. Shirai. Space division for ray tracing in CSG. *IEEE Computer Graphics and Applications*, 6(4):28–34, 1986.

[168]K. Yamaguchi, T.L. Kunii, K. Fujimura, and H. Toriya. Octree-related data structure and algorithms. *IEEE Computer Graphics and Applications*, 4(1):53–59, 1984.

[169]M.A. Yerry and M.S. Shephard. Automatic mesh generation for three-dimensional solids. *Computers and Structures*, 20(1–3):31–39, 1985.

[170]N Yoshimura. P-version of fem. *CAD & CIM*, 4(2):61–63, 1990. (in Japanese).

Index